# Lecture Notes in Control and Information Sciences

Edited by A. V. Balakrishnan and M. Thoma

Vol. 1: Distributed Parameter Systems: Modelling
and Identification
Proceedings of the IFIP Working Conference,
Rome, Italy, June 21–26, 1976
Edited by A. Ruberti
V, 458 pages. 1978

Vol. 2: New Trends in Systems Analysis
International Symposium, Versailles,
December 13–17, 1976
Edited by A. Bensoussan and J. L. Lions
VII, 759 pages. 1977

Vol. 3: Differential Games and Applications
Proceedings of a Workshop, Enschede, Netherlands,
March 16–25, 1977
Edited by P. Hagedorn, H. W. Knobloch, and G. J. Olsder
XII, 236 pages. 1977

Vol. 4: M. A. Crane, A. J. Lemoine
An Introduction to the Regenerative Method for
Simulation Analysis
VII, 111 pages. 1977

Vol. 5: David J. Clements, Brian D. O. Anderson
Singular Optimal Control: The Linear Quadratic Problem
V, 93 pages. 1978

Vol. 6: Optimization Techniques
Proceedings of the 8th IFIP Conference on Optimi-
zation Techniques, Würzburg, September 5–9, 1977
Part 1
Edited by J. Stoer
XIII, 528 pages. 1978

Vol. 7: Optimization Techniques
Proceedings of the 8th IFIP Conference on Optimi-
zation Techniques, Würzburg, September 5–9, 1977
Part 2
Edited by J. Stoer
XIII, 512 pages. 1978

Vol. 8: R. F. Curtain, A. J. Pritchard
Infinite Dimensional Linear Systems Theory
VII, 298 pages. 1978

Vol. 9: Y. M. El-Fattah, C. Foulard
Learning Systems:
Decision, Simulation, and Control
VII, 119 pages. 1978

Vol. 10: J. M. Maciejowski
The Modelling of Systems with Small Observation Sets
VII, 241 pages. 1978

Vol. 11: Y. Sewaragi, T. Soeda, S. Omatu
Modelling, Estimation, and Their Applications for
Distributed Parameter Systems
VI, 269 pages. 1978

Vol. 12: I. Postlethwaite, A. G. J. McFarlane
A Complex Variable Approach to the Analysis of
Linear Multivariable Feedback Systems
IV, 177 pages. 1979

Vol. 13: E. D. Sontag
Polynomial Response Maps
VIII, 168 pages. 1979

Vol. 14: International Symposium on Systems
Optimization and Analysis
Rocquentcourt, December 11–13, 1978;
IRIA LABORIA
Edited by A. Bensoussan and J. Lions
VIII, 332 pages. 1979

Vol. 15: Semi-Infinite Programming
Proceedings of a Workshop, Bad Honnef,
August 30 – September 1, 1978
V, 180 pages. 1979

Vol. 16: Stochastic Control Theory
and Stochastic Differential Systems
Proceedings of a Workshop of the „Sonder-
forschungsbereich 72 der Deutschen Forschungs-
gemeinschaft an der Universität Bonn"
which took place in January 1979 at Bad Honnef
VIII, 615 pages. 1979

Vol. 17: O. I. Franksen, P. Falster, F. J. Evans
Qualitative Aspects of Large Scale Systems
Developing Design Rules Using APL
XII, 119 pages. 1979

Vol. 18: Modelling and Optimization of Complex
Systems
Proceedings of the IFIP-TC 7 Working Conference
Novosibirsk, USSR, 3–9 July, 1978
Edited by G. I. Marchuk
VI, 293 pages. 1979

Vol. 19: Global and Large Scale System Models
Proceedings of the Center for Advanced Studies (CAS)
International Summer Seminar
Dubrovnik, Yugoslavia, August 21–26, 1978
Edited by B. Lazarević
VIII, 232 pages. 1979

Vol. 20: B. Egardt
Stability of Adaptive Controllers
V, 158 pages. 1979

Vol. 21: Martin B. Zarrop
Optimal Experiment Design for
Dynamic System Identification
X, 197 pages. 1979

*For further listing of published volumes please turn over to inside of back cover.*

# Lecture Notes in Control and Information Sciences

Edited by A.V. Balakrishnan and M. Thoma

## 35

## Global Modelling

Proceedings of the IFIP-WG 7/1 Working Conference
Dubrovnik, Yugoslavia, Sept. 1–5, 1980

Edited by S. Krčevinac

Springer-Verlag
Berlin Heidelberg GmbH 1981

ISBN 978-3-540-11037-8    ISBN 978-3-540-38560-8 (eBook)
DOI 10.1007/978-3-540-38560-8

# PREFACE

This volume contains a collection of papers presented at the IFIP
Working Conference on Global Modelling which was held from September
1-5, 1980 in Dubrovnik, Yugoslavia. The Conference was sponsored by
the IFIP TC-7 - System Modelling and Optimization and organized by
the Faculty of Organizational Sciences, University of Belgrade and
ETAN (Yugoslav Committee for Electronics and Automation).

The Conference had as an objective to discuss the methodological and
user-oriented problems of Global Models and to provide for an opportunity of better communication among modellers and users. We believe
that this objective was, at least partially, met.

We are indebted to the members of the International Program Committee
for their help in finalizing the program of the Conference and for
choosing the papers. In particular, the help of Professor R.F. Drenick,
Chairman of the International Program Committee, Professor R. Tomović,
Chairman of the Conference and Professor B. Lazarević, Chairman of the
National Organizing Committee, is gratefully acknowledged.

We also want to thank Mr. Bratislav Petrovic for editorial assistance
and Mrs. Brenda Marjanović for her reliability and competence in typing
the proceedings.


June, 1981
Belgrade                                   S. Krčevinac

# CONTENTS

R. Tomović
Top-Down Approach to Global Dynamic Modelling ................ 1

Z. Zarić
Experience with Energy System Modelling in Serbia ........... 5

Y.R. Cho and E.D. Gahan
The UNIDO World Industry Co-operation Model ................ 18

S.I. Gass and S.C. Parikh
Credible Baseline Analysis for Multi-Model Public Policy
Studies ..................................................... 34

H.P. Smit, R.P. Vos and H.J.W. Weyland
Medium- and Long-Term Models for the ESCAP Region .......... 57

H. Bossel
Modelling Policy Consequences and Evaluation Processes
Using the "DEDUC" Nonnumerical Program System .............. 101

A.H. Zemanian
Economic Models of Periodic Marketing Systems .............. 119

Y. Zahavi
A New Urban Travel Model ................................... 130

P.M. Allen
Modelling the Self-Organization of Human Systems ........... 139

S.K. Gehrecke
The Impact of Energy and Environmental Policy on the
Design of Energy Models .................................... 173

N. Fergany
Treatment of the Arab Region in Global Models .............. 186

G. Bruckmann
IIASA's Role in Global Modeling ............................ 192

V.B. Bajić and B.J. Petrović
Quasi-Models of Price Evolution and Their Qualitative
Properties ................................................. 196

G.G. Pirogov
A Model of Regional Interactions Considering Energy Deficits.. 206

M. Rajkov, S. Andrić, Z. Šišarica and Lj. Rajin
Contribution to the Simulation Modelling of an Economic
System ..................................................... 218

IFIP WORKING CONFERENCE ON GLOBAL MODELLING
Dubrovnik, Yugoslavia
September 1-5, 1980


LIST OF PARTICIPANTS

| | |
|---|---|
| BAJIĆ Vladimir | PTT Educational Center, Zdravka Čelara 16, Belgrade, Yugoslavia. |
| BATALOV V. | Institut "Boris Kidrič" - Vinča, Belgrade, Yugoslavia. |
| BINGULAC S. | Institut "Mihailo Pupin", Volgina 15, Belgrade, Yugoslavia. |
| BRUCKMANN Gerhart | University Prof. Dr., IIASA, A-2361 Laxenburg, Austria. |
| BOSSEL Hartmut | University of Kassel, Dept. of Matm. - Envir. Systems Analysis, D35 Kassel, F.R.G. |
| ĆIRIĆ Vidojko | Faculty of Organizational Sciences, Oslobodjenja 1, Belgrade, Yugoslavia. |
| ĆIRIĆ Duško | Gradski Zavod za informatiku, Tiršova 1, Belgrade, Yugoslavia. |
| DENEŠ Oto | Yugoslav Culture Center in Paris, 123 Rue St Martin, Paris, France. |
| DRENICK Rudolf F. | Polytechnic Institute of New York, 333 Jay Str. Brooklyn NY 11201, USA. |
| ELSHAFEI A.N. | London School of Economics, London, England. |
| ERSU Enis | Fachgebiet Regelsystemtheorie, Technische Hochschule Darmstadt, Germany. |
| FERGANY Nader | Arab Planning Institute, Kuwait, Kuwait. |
| GAHAN Eoin | UNIDO Vienna, Austria. |
| GASS Saul I. | University of Maryland, College Park, Maryland, USA. |
| GEHRECKE Siegfried | Fakultet organizacionih nauka, Oslobodjenja 1, Belgrade, Yugoslavia. |
| HAN Stjepan | Blvr. Maršala Tita 6/IV, N. Sad, Yugoslavia. |
| HUBAND Frank | 1204 N. Stafford, Arlington, Va, USA. |
| ILIĆ Dragan | Narodna banka Crne Gore, Titograd, Yugoslavia. |
| JAKŠIĆ-LEVI Maja | Fakultet organizacionih nauka, Oslobodjenja 1, Belgrade, Yugoslavia. |
| JAKŠIĆ Dušan | Univerzitet Novi Sad, Veljka Vlahovića 3, Novi Sad, Yugoslavia. |
| JOVANOVIĆ Savo | Institut "Boris Kidrič", Vinča, Belgrade, Yugoslavia. |
| JOVANOVIĆ Vladan | Fakultet organizacionih nauka, Oslobodjenja 1, Belgrade, Yugoslavia. |
| KONVALINKA Ira | Institut "Boris Kidrič", Vinča, Belgrade, Yugoslavia |
| KOSTIĆ Ljiljana | "Boris Kidrič"-Vinča, Belgrade, Yugoslavia. |

| | |
|---|---|
| KRČEVINAC Slobodan | Faculty of Organizational Sciences, Belgrade, Yugoslavia. |
| LAZAREVIĆ Branko | Fakultet organizacionih nauka, Oslobodjenja 1, Belgrade, Yugoslavia. |
| KOVILJKO Lovre | Ekonomski institut, Subotica, Yugoslavia. |
| MESAROVIĆ Mihajlo | Case Western Reserve University, Cleveland, Ohio, USA. |
| MILENKOVIĆ Čedomir | Institut "Boris Kidrič" Belgrade, Yugoslavia. |
| MORA Andraš | Izvršno Veće SAP Vojvodine, N.Sad, Yugoslavia. |
| OBRADOVIĆ Danilo | FTN, Veljka Vlahovića 3, N. Sad, Yugoslavia. |
| PETRIĆ Jovan | Fakultet organizacionih nauka, Oslobodjenja 1, Belgrade, Yugoslavia. |
| PIROGOV Grigorii G. | Moskow, Ryleev Street 14, USSR. |
| PETROVIĆ Mirko | Fakultet organizacionih nauka, Oslobodjenja 1, Belgrade, Yugoslavia. |
| PETROVIĆ Bratislav | Fakultet organizacionih nauka, Oslobodjenja 1, Belgrade, Yugoslavia. |
| POPOVIĆ Milan | Beogradska 23, Titograd, Yugoslavia. |
| RAJKOV Miloš | Fakultet organizacionih nauka, Oslobodjenja 1, Belgrade, Yugoslavia. |
| RISTIĆ Slobodan | Administration for International Scientific, Educational, Cultural and Technical Cooperation of the Socialist Republic of Serbia, Nemanjina 22-III, Belgrade, Yugoslavia. |
| SMIT Hidde P. | Free University, Department of Economics, Amsterdam, Netherlands. |
| ŠINKA Milan | Buda Tomovića 5, Belgrade, Yugoslavia. |
| ŠOMODI Šandor | Institut za organizaciju poslovanja, Subotica, Yugoslavia. |
| STOJANOVIĆ Zoran | Novi Sad Univerzitet, Veljka Vlahovića 3, Novi Sad, Yugoslavia. |
| ŠPOLJARIĆ Vesna | Fakultet organizacionih nauka, Oslobodjenja 1, Belgrade, Yugoslavia. |
| TOJAGIĆ Slobodan | Institut za teh. mesa, mleka, ulja, masti, voća i povrća, Tehnološki fakultet, Novi Sad, Rumenačka 103, Novi Sad, Yugoslavia. |
| TOMOVIĆ Rajko | Elektrotehnički fakultet, Bulevar revolucije 73, Belgrade, Yugoslavia. |
| VUKOVIĆ Nahod | Fakultet organizacionih nauka, Oslobodjenja 1, Belgrade, Yugoslavia. |
| VUKIĆEVIĆ Pavle | Galšići 54, Titograd, Yugoslavia. |
| VUKASOJEVIĆ Radomir | Mašinski fakultet, Titograd, Yugoslavia. |
| WEYLAND Hermine J.W. | Vrye Universiteit, Dept of Economics, Amsterdam, Holland. |
| ZARIĆ Zoran | Institut "Boris Kidrič", Belgrade, Yugoslavia. |
| ZEMANIAN Armen H. | State University of New York at Stony Brook, Stony Brook N.Y., USA. |

ZELENOVIĆ Dragutin      Novi Sad, Veljka Vlahovića 3, Novi Sad, Yugoslavia.

ZAHAVI Yacov      7304 Broxburn Ct, Bethesda MD. 20034, USA.

# TOP-DOWN APPROACH TO GLOBAL DYNAMIC MODELLING

Rajko Tomović

Faculty of Electrical Engineering, Belgrade, Yugoslavia

## Introduction

The publicity given to world development models was certainly
out of proportion compared to the state of the art of the large system
theory. Today, there is almost a general agreement that the formal
theory of large systems cannot be considered as a mere extension of
the state space approach and computer simulation methods which are so
helpful in studying mechanical and engineering problems. Dominant
issues in large system studies are of a structural nature such as vertical and horizontal (subsystem) decomposition, not to speak of optimization in the presence of conflicting criterion functions. The final
consequence of the above theoretical situation in large system theory
is, however, quite practical. The user, for which the global model is
ultimately built, has no definite guarantees as to the validity of
predictions and conclusions derived from computer simulation.

In this respect, the users of global models in developing
countries must take special precautions. Namely, models of "soft"
dynamic systems do reflect necessarily in a more or less implicit way
many socio-economic assumptions and national development patterns which
are simply not true in the environment of the actual user. Consequently, if the user´s confidence in global models has to be increased,
it is essential to understand the overall effects of the environment
on the validity of global models.

This inherent shortcoming of global modelling can be overcome in several ways. In the first place, dynamic models of soft systems must be made transparent. In other words, the impact of the
socio-economic environment and development patterns on the derivation
of the global model must be made explicit and brought clearly to the
attention of the user. This requirement is becoming today an essential part of any high-quality model building in the field of soft
large systems.

In our opinion, an even more powerful approach helping to overcome the abovementioned shortcomings of global models has been developed by the relevant multidisciplinary group in Belgrade (1). The approach in question has been given the name Top-Down Design. The basic facts about the top-down approach to global modelling will be presented in the next section.

## Top-Down Design

As mentioned above, the current practice in supplying the global model and the corresponding software reflects the prevailing attitude that the modelling results are either insensitive to the socio--economic environment of the user or that the user will essentially follow the conceptual approach and the national development pattern of the model builder. However, this is simply not correct. In order to make transparent the role of deep layers hidden behind mathematical relations and software structure of global models, the usual methodology of starting at the level of computer simulation has been turned upside down. Before actually assessing the value of global modelling, an extensive study of the multilayer conceptual structure leading to formal description of the dynamic system has been undertaken. Special attention was paid to the effect of the socio-economic environment and alternative development patterns upon the quantitative form of the global model. The Yugoslav environment and the development pattern of a typically underdeveloped country to the current stage of a medium-developed society offer, no doubt, a great challenge.

In our case, the Top-Down Approach begins with general social and political considerations which ultimately reflect upon the quantitative level and the model structure. Evidently, such basic features of a socio-political system as market or state controlled pricing, investment policies, wealth ownership and distribution among social groups, self-government oriented society, etc., do affect the basic premises on which the quantitative relations and econometric expressions are derived. Unfortunately, extensive comparative studies of the interaction between the socio-political principles and quantitative relations related to global modelling are simply not available.

Historical and cultural environments are another major factor to be entered in global model building. For instance, the mobility and availability of work force are taken for granted in many economic considerations. However, in economies constituted from highly heterogeneous regions in terms of historical, cultural and religious evolution, the mobility of the work force is not just a matter of economic policy

It is reasonable to assume that when approaching the quantitative level of the global model more and more things will be commonly shared. In principle, this is true but even here a considerable measure of caution is needed. For instance, the efficiency of supply-demand regulation using the pricing mechanism will be greatly reduced if non--renewable raw materials are involved or if the overhead versus direct cost ratio is drastically distorted in favour of other types of expenses. The amount of goods to be exported from a country cannot be, in fact, predicted correctly by simple balancing equations. The degree of national independence, overall priorities in national development, high or low probability of being exposed to agression - all these factors are involved in national export-import policies.

The above considerations are mentioned here in order to point out the sensitivity of global models to the overall environmental conditions surrounding the model user. As a matter of fact, the idea of checking the sensitivity of global models in order to derive conclusions on their reliability occurred quite early on in the evolution of this branch of system studies. However, the research was associated with sensitivity studies to parameter variations of the model. The attitude taken by the Belgrade Group on global modelling is essentially different. Our main concern has been to analyze the model´s sensitivity to environmental changes which has not been done in a systematic way so far. To be quite precise, the validity of global models in different environmental conditions has been checked but in a highly undesirable way: by the users themselves. As is known, the high rejection ratio of any model by the final user is not the most constructive way to explore the sensitivity of dynamic models to environmental changes.

Conclusion

The sensitivity analysis of global models undertakin in Yugoslavia leads to some important implications. In the first place, the direct transfer of global models into different environments is hardly productive. Global models must be developed as a set of choices derived from different starting positions. Consequently, the activity in global modelling must be a decentralized world effort. If it comes down to the transfer of knowledge in this area, then the process should start by sharing the *data bases* needed for derivation of global models and software products. The assessment of the validity of the transferred model, modifications and adaptation to local needs must be generated by expert teams knowledgable in both local problems and system theory. Further evolution of the Top-Down Approach developed in Yugoslavia has lead to the following decision. A special multidisciplinary study group has been formed at the Faculty of Organizational Sciences, Belgrade, under the sponsorship of industry and planning boards. The main task of the Group shall be to implement the available software for global modelling on local computers and proceed, on the basis of previous critical studies on model sensitivity, to the assessment of existing results and generate appropriate modifications. In short, we believe that global modelling may eventually become a tool for improved decision making in industry and government but in order to reach that stage of reliability a great deal of academic and applied multidisciplinary research of a decentralized nature must be undertaken.

Reference

1. Developing Countries and Progress Perspectives, S. Radoman, Editor, Publishing House "Communist", Belgrade, 1979.

# EXPERIENCE WITH ENERGY SYSTEM MODELLING IN SERBIA

Zoran Zarić

Boris Kidrič Institute, Belgrade, Yugoslavia

## Introduction

Serbia is one of the six federal republics of Yugoslavia. Without the two autonomous provinces, Serbia has about 5.5 million inhabitants. It seems, therefore, that there is some discrepancy between the topic of this paper and the general topic of the Conference on Global Modelling. Let us, however, look at the principal characteristics of the energy economy in Serbia. Essentially, these are:

- annual primary energy consumption per capita slightly below the global average of about 2 kWyears, increasing lately by about two per cent annually,
- energy demand characterized by presumably low end use efficiency and an unfavourable structure based on more than 50 per cent imported oil and coke and on about 20 per cent electricity,
- energy supply relying to a sizable extent on imported fuels as well as on low grade domestic lignite, with very little upgrading of the lignite which is mostly used in conventional power plants with modest efficiency and partly imported technology,
- known energy resources principally in surface mined, low grade lignite amounting to some 250 kWyears per capita. Indicated resources in uranium, oil shale and geothermal energy in an undetermined quantity. Research and development efforts in the development of new technologies and unconventional energy sources are very modest.

In short, the energy economy in Serbia is characterized by limited resources, a large percentage of imported fuel, conventional and partly imported technology, an unfavourable and wasteful consumption structure, modest research efforts and modest experience with energy modelling and planning. Does this sound familiar? It should as these characteristics are common to most of the developing countries which comprize three-quarters of mankind at present and even more in the future.

Therefore, it seems that there is a reasonable ground to assume that the energy system modelling experience in Serbia might be of more general interest and of some use on a global scale.

## A Case for Energy Modelling in Developing Countries

It is interesting to look at the prospects of future energy supply in most of the countries of the world belonging to the category of developing countries. Although predictions of future energy demand and supply are necessarily uncertain, they are indispensable in order to be prepared for the future, and evaluate existing possibilities. A number of studies by international organizations, such as the World Energy Conference, IIASA and others, have lately been concerned with such predictions, a partial summary of which is given in (1). Figures 1, 2 and 3 sum up the principal findings. In these figures, the world is roughly divided into two parts; into developing and developed countries. In Fig. 1, predictions of the probable population growth are given for the world (1), for the developing countries (2) and for the developed countries (3). It can be seen that by the end of the next century close to 85 per cent of the global population will be living in those countries which are at present in development. Corresponding estimates on the per capita energy consumption growth are presented in Fig. 2. It is interesting to note that in these estimates growth in the developing countries is more modest than in the developed countries well into the next century. In spite of this, energy demand in developing countries will surpass energy demand in developed countries in the first half of the next century, as is seen from Fig. 3, resulting from Figs. 1 and 2.

Estimates on energy supply possibilities from conventional sources (fossil fuels, nuclear convertors and renewable resources like hydropower, firewood and wastes) in developed and developing countries are shown in Fig. 4. It is seen from these estimates that energy supply from fossil fuels in developing countries will start to decline early in the next century. This is a consequence of the fact that most of the coal resources are situated in developed countries. Energy supply and demand are matched in Fig. 3. It is seen that at the present there is a surplus of energy supply over demand in developing countries as the result of oil resources being located mostly in these countries. This surplus is transferred to developed countries to cover the energy deficit. It is estimated that this transfer will continue at the beginning of the next century as the oil exporting countries would be in need of hard currency for their development. This is represented by hatched areas in Fig. 3.

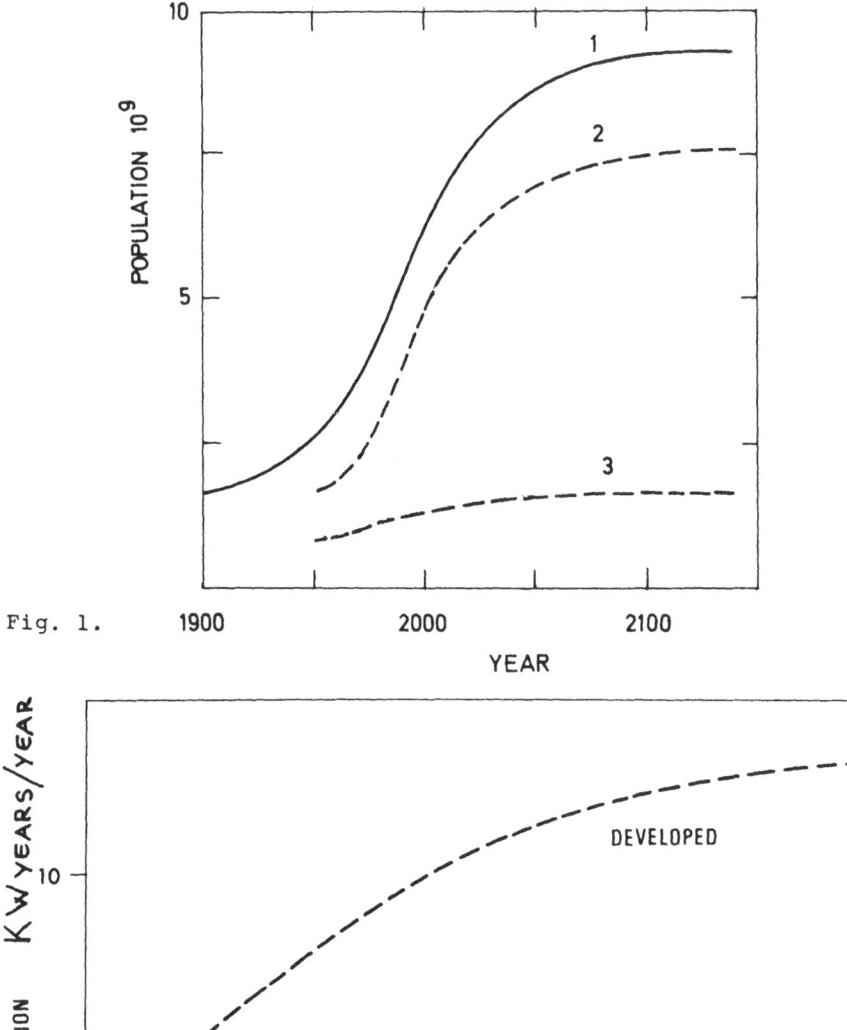

Fig. 1.

Fig. 2.

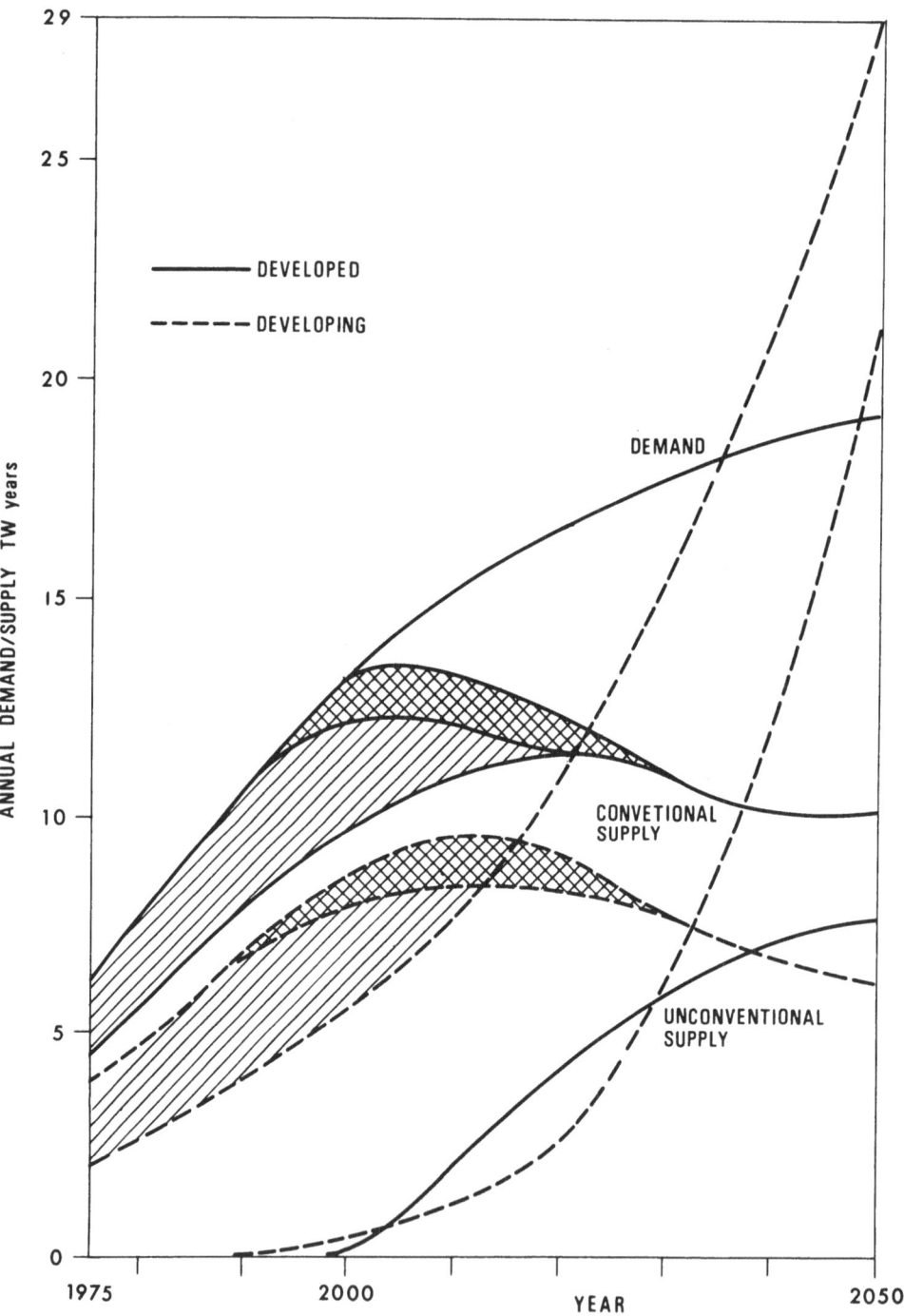

Fig. 3.

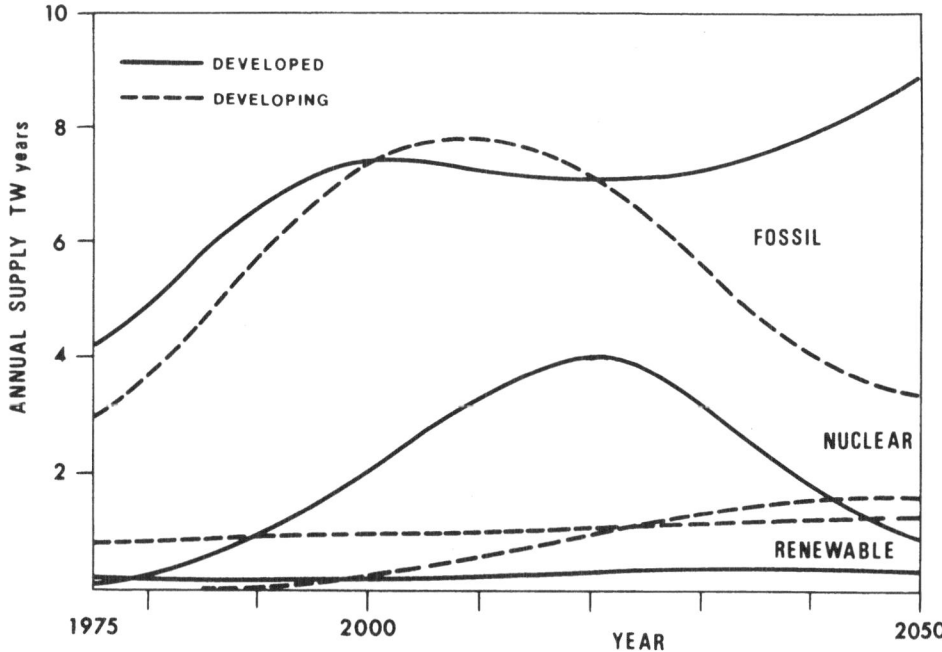

Fig. 4.

Table 1  Fuel costs and specific investments for different electric
power plants /2/.

| Power plant | Fuel costs mills/kWh | Specific invest-ments $/kW |
|---|---|---|
| Oil | 70 | 420 |
| Coal | 25 | 550 |
| Nuclear (PWR) | 5 | 1 000 |
| Nuclear breeders | 2 | 1 500 |
| Hydroelectric | 0 | 3 200 |
| Solar electric | 0 | 14 000 |
| Geothermal electric | 0 | 15 500 |

The gap between supply and demand will start to form according to these estimates, as early as by the end of this century. This gap has to be filled by unconventional energy sources - nuclear and thermonuclear breeders, solar and geothermal energy. From Fig. 3 it can be seen that unconventional sources have to be introduced in developing countries probably earlier than in the developed countries. It is also apparent that the market penetration of these sources in developing countries should be very fast and that the unconventional sources are estimated to become principal energy sources in these countries as early as 30 to 40 years from now.

Now, unconventional energy sources differ very much in technology, ranging from soft solar energy utilization to highly sophisticated fusion reactors. However, they all have some characteristics in common:

- fuel costs are negligible or even non-existent. On the contrary, investment costs become very high, much higher than with conventional sources. This can be seen from Table 1, made on the basis of the data from (2).
- a sizable part of investment costs goes on know-how and quality assurance, much more so than with conventional sources.
- most of the technologies on which the exploitation of unconventional sources are based are in various stages of development, some of them even in the laboratory stage. Research and development costs are very high and leading times very long. It is estimated that 20 to 30 years are needed for a technology from the laboratory stage, or proven scientific feasibility, to a full scale commercial plant. For a technology to penetrate the market to a sizable extent in the first quarter of the next century, the time for decision is now.
- each of the energy sources which could satisfy future energy demand, like nuclear or thermonuclear breeders, solar energy or geothermal energy, could be developed in a number of directions, each of them having advantages and disadvantages. Each of these directions is based on a particular technology requiring considerable time, many and human efforts to be developed. For instance, solar energy could be transferred into heat, electricity or fuels. Each of these possibilities could be developed in half a dozen variations, each based on a particular technology. Electricity could be obtained through thermo--mechanical or through photovoltaic conversion characterized by completely different technologies. Further on, photovoltaic convertors could be based on a number of semi-conductors, could be with or without concentrators, etc.

To summarize: future energy supply will not any more be based on the wealth buried in the ground but on human know-how, skills and efforts. None of the developing countries is rich enough to rely on imported technology in the future. The development of appropriate technologies has therefore to be initiated very soon and supported by available funds. The funds are, however, scarce in such countries so that it is necessary to be selective and concentrate on a few well chosen technologies. The decision is of considerable importance so that it has to be based on all available information as well as on scientific grounds.

Energy system modelling is the only available alternative for the serious evaluation of advantages and disadvantages of different technologies for the future. This in itself represents a strong case for energy modelling in developing countries. There are additional gains, also. As mentioned in the Introduction, energy economy in developing countries is not usually in a good shape. In this respect, much could be gained from a systematic analysis of the energy economy based on energy modelling. In most of the developing countries there is also the urgent problem of substituting imported fuels by synthetic products based on domestic resources, like gas or liquid fuels from coal or biomass. Here, also, there is always a number of possible routes and a decision has to be made to follow only some of them. Again, energy system modelling based on all available information is indispensable.

## Models and Modelling

There is usually very limited experience, if any, with energy system modelling in most of the developing countries. In addition, there is usually a lack of adequate information on which modelling has to be based. A wide variety of more or less sophisticated energy models is available in developed countries where there is also good experience with their application. A number of research centres in the USA, USSR and West Europe, like Brookhaven, Irkutsk, Jullich, Laxenburg and others, have been engaged for a number of years in developing energy system models and applying them to the analysis of energy economy on a national or a global scale. Some of these centres would be ready to help in the energy economy analysis of particular developing countries and, in fact, a number of such studies have been made. Energy economy is, however, a vital part of the economy of a country and has to be analyzed by people from the country who have adequate experience and know the situation in that country. Besides, there is no point in analyzing energy economy outside the country as energy models are efficient tools which have to be applied constantly using improved information.

Alternatively, a number of models available in developed countries could be obtained on request with an assurance of adequate training in their usage. This opportunity is very useful but the choice of a particular model still has to be made.

Available models range from relatively simple energy sector models to systems of sophisticated multisectorial models. The indispensable part is the energy supply model of the kind shown schematically in Fig. 5, in which energy flow paths are represented through all conversion steps, from resources to end uses. Useful energy demand and available energy technologies are inputs to the model. Consumption of primary resources, necessary investments, material and labour, energy costs and environmental impacts are outlets. Models can be of a simulation kind of the "what if?" type. Models can be provided with an optimization procedure, in which case an optimum structure is obtained. Models can be static (i.e., for a given year) or dynamic (for a given period in the future), and differ in the level of aggregation. In an optimization model, an optimum structure of technologies and interfuel substitution for a given goal function which has to be provided exogenously is obtained. Usually a single goal function, as for instance energy costs, is insufficient so that a multiobjective analysis has to be performed.

At the other extreme, a system of models is employed in the sense described in Fig. 6. A macroeconomy model defines energy demand which is used as an input in the energy supply model. An economic impact model uses the output from the supply model to make a link with the macroeconomy model. Submodels for resources, technologies, processes, environmental impact, etc., could also be linked to the system. Such a system of models is usually the end result of the development of energy modelling in a country in a number of years whereas the models are constantly improved and increase in complexity. In the end, models are so large that essential relationships are obscured, no one knows the whole model, outputs are very difficult to interpret, data requirements are huge, the complexity of the models is not justified by the available data quality and analysis requires a long time, large computers and running costs. Therefore, the trend, at least in the USA, is to carry out up to 90 per cent of analysis using simple sectorial models and employing sophisticated systems of models only in cases where great detail is required (3).

The principal problem is adequate data, as the quality of the analysis depends on their quality. Historical data on energy supply and demand are necessary for analysis of the present energy economy as well as for estimates of future trends. Data collected through official

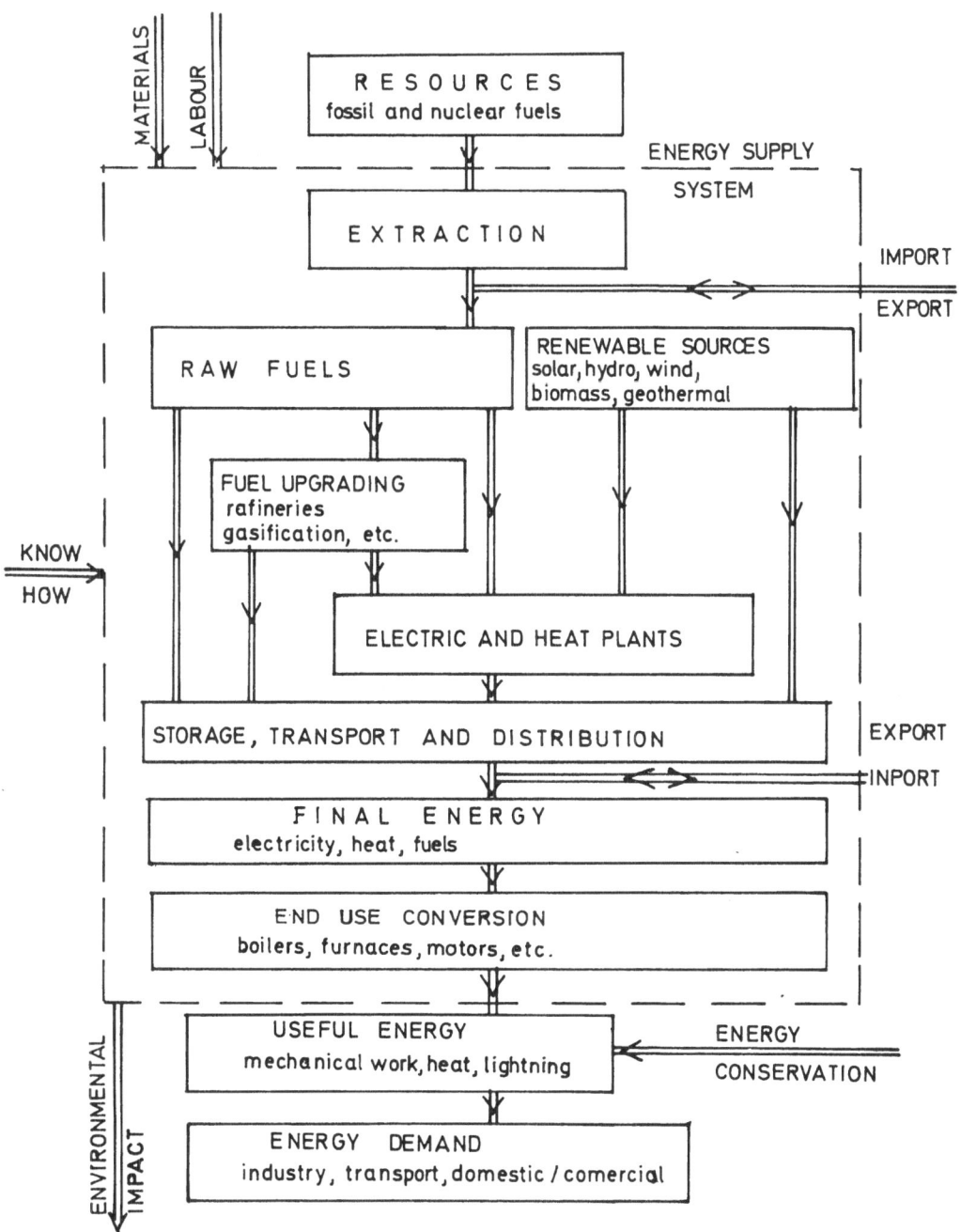

Fig. 5.

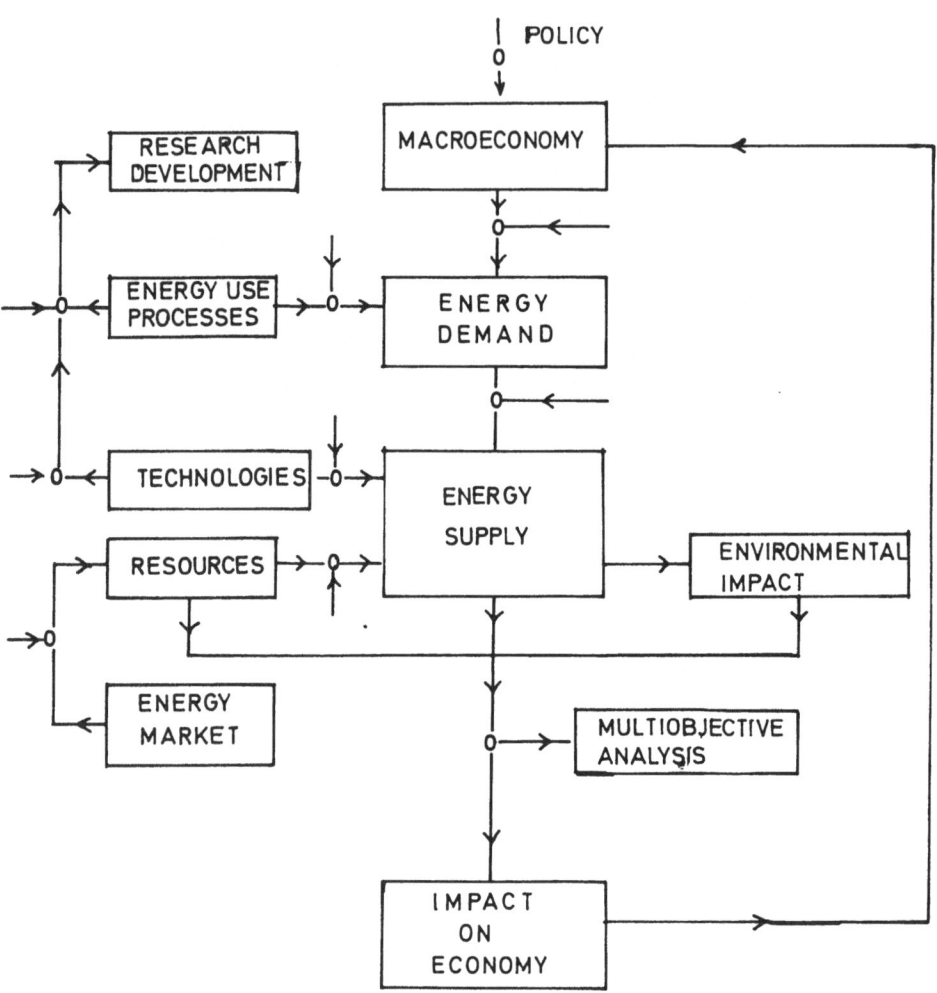

Fig. 6.

statistics are usually inadequate, especially in developing countries. Data are unreliable, handled manually and stored in the form of tables. Better organization of data collection and a sort of energy information system in the form of a computerized data book is necessary. Energy models have a significant role in the organization of the information system defining which data are essential for analysis. Anyway, the build-up of an energy information system is a process requiring a number of years and until it is successfully implemented there is no use for sophisticated energy models. It has to be noted that economic characteristics are as essential as energy balances and these are as a rule more difficult to obtain.

In addition to historical data on energy supply and consumption, techno-economic data on new prospective technologies of energy conversions and unconventional energy sources are needed. The reliability of such data is always questionable so that they have to be collected from various supposedly objective sources and very critically reviewed. This requires close contacts with various institutions in industrially developed countries, developing these technologies, as well as a competent group of domestic specialists knowledgeable in different fields. A well organized effort in this direction is necessary.

## Experience in Serbia

Some experience with energy modelling has existed for some time in Yugoslavia. The group of Professor Požar in Zagreb developed a number of energy supply models of both simulation and optimization type (4). The problems encountered were mostly in the area of data collection and availability. In 1977, a multi-institutional project started in Serbia with the aim of making a systematic effort in energy modelling. The project included a number of scientific institutions in the fields of mining, mechanical, electrical and traffic engineering, nuclear energy, agriculture and economy. The project is in its final stage of realization (5).

A task force, composed of various specialists, is engaged to systematically collect and critically review the existing historical data on energy supply and consumption for the period of at least 10 to 20 years. Official statistics were taken as the base but it came about that these were not always appropriate. End use data belong to this category, especially in domestic and commercial sectors where the break down of consumption was unsatisfactory and had to be obtained by organized research. Data on end use efficiencies were also very scarce, and unreliable. Additional efforts were therefore necessary to establish a set of data needed for modelling. Another task force was engaged on collecting energy prices, investment costs and other data needed for economic evaluations. These data were even more difficult to collect and evaluate critically. Yet another task group was involved in collecting data on environmental impact, mainly in the form of unit emissions. These data are mostly based on estimates.

It was decided early on that at this stage modelling should be restricted to the energy sector alone. Only after gaining some experience with the energy supply models could one go further and couple them with some macroeconomic models in order to establish an energy--economy link. This then means that energy demand in the future has to be supplied to the model exogeneously, with the aid of econometric relations based on historical data. However, since interfuel substitution should be one of the goals of optimization energy demand has to be given in the form of useful energy. In making projections in the future, data are needed on new energy conversion technologies and new energy sources. A special task force composed of various specialists was therefore engaged in the collection and evaluation of technical and economic data of different prospective technologies. Contacts with groups in industrially developed countries engaged in similar evaluations was very helpful.

A group in the Boris Kidrič Institute is involved in developing an energy supply model to be used in evaluating the present energy economy as well as different policies for energy supply in the future. The group established close contacts with energy modelling centres in Brookhaven, USA, Jüllich, FR Germany and elsewhere. Detailed information was obtained on several existing models developed and used in these centres. Finally, the Brookhaven Energy System Optimization Model (BESOM) was chosen for the implementation of energy economy in Serbia at this stage. The Brookhaven National Laboratory was very helpful in providing the programme as well as the people engaged in its development which greatly facilitated the programme's implementation. The programme is installed at a CYBER computer in Yugoslavia.

BESOM is a linear programming, static, energy supply model. This means that there are other more sophisticated models available. However, most of these sophisticated models were developed on the basis of BESOM. Furthermore, the quality of data available at this stage does not justify the use of more elaborate models. BESOM is readily accessible, which is very important as a number of adjustments had to be made in it in order to comply with the structure of energy economy in Serbia. This allowed a number of people to become well acquainted with BESOM so that access to more elaborate models in the future is greatly facilitated. A number of analyses have been made to date of the present situation as well as of some scenarios for developments in the future. Since there are no links to a macroeconomy model, multiobjective analysis is applied, for which BESOM is very well adjusted. In addition to energy costs, supply security, investment costs, environmental impact and conservation of resources were employed as goal functions.

## References

1. Zarić, Z., Energy Sources for the Future, UNESCO (in print).
2. Energy Technology Data Handbook, Vol. I Conversion Technologies, KFA Jüllich, Jül-Spez-70 (1980).
3. Brock, H.W.; Nesbitt, D.M., Large Scale Energy Planning Models, A Methodological Analysis, Stanford Research Institute (1977).
4. Požar, H.; Vuk, B., Matematički model za optimizaciju energetske strukture, Zajednica Jugoslovenske elektroprivrede, Zagreb (1979).
5. Zarić, Z., Modeliranje sistema energetike i primena na energetiku SR Srbije. Zbornik Energetika Srbije 80. 865/886 (1980).

THE UNIDO WORLD INDUSTRY CO-OPERATION MODEL

Y.R. Cho and E.D. Gahan

United Nations Industrial Development Organization,
Vienna, Austria

## 1. Introduction

This paper is intended to give an outline of the UNIDO World
Industry Co-operation Model.  In this rather long title each of the
words has been chosen in an attempt to capture the basic ideas behind
it.

The word "world" has been chosen to emphasize that the model
deals with the "world" rather than "global" economy in terms of its
component nations.  This means that the intention is to have, in this
economic model, a detailed consideration of each country, and this is
because it is ultimately individual countries who are the policy makers
for their economies.  Since a primary use of the model is to examine
policies, there is no point in only having world aggregate or regional
detail alone in the model, because the regional policy, derived from
the model, would still have to be interpreted in terms of national
policies.

The use of the word "industry" in the title is intended to
emphasize not only UNIDO involvement, but also the fact that, in the
construction of the model, considerable attention has been paid to
allowing a detailed examination of the manufacturing sector both at the
level of international trade and also in terms of national planning.

"Co-operation" is used to highlight several features of the
model.  Firstly, the model incorporates and examines the agreed development goals of the international community, in particular the Lima target, which are to be achieved by rational and collective action on the
part of the world nations.  Secondly, the model itself is intended to
be a mechanism which allows national planners from different countries

Of many who have contributed to the project work, special mention must
be made of J. Sivak (Computer Centre of the National Planning Office,
Hungary) in connection with the overall model structure, and M. Oettl,
K. Mauler and P. Ahammer (Wirtschaftsuniversität, Vienna) in connection with the LIDO Model.

to examine their economic strategies, in relation to those of other
countries.

The purpose of the model may be summarized as follows: it is
intended to be a system whereby individual country plans and projections
may be examined and adjusted in a world context. Thus it provides for
the reconciliation of national macroeconomic strategies in the light of
those of other countries and of regional and global goals and scenarios.
It is intended to show also the detailed implications of such reconciled strategies in terms of bilateral trade and its commodity composition, and this analysis is then further extended to assess the national
sectoral implications. The model system is therefore interactive in
its concept and its implementation, and can be seen as an educational
device, one which, it is hoped, may be useful not only for individual
country analysis, but ultimately also as an information tool in
international negotiation.

The next section of this paper gives an account of the various
modules of the inner and outer layers of the system. Section 3 concentrates on an important sub-model, the LIDO Model. This is a small world
model built to examine the implications of the Lima target in the year
2000. It produces results for intermediate years also, and thus
supplies constraints to the UNIDO model system as a whole, because it
shows the kind of area into which national projections have to fit if
the Lima targets are to be achieved. As well as this, it is used in
UNIDO to carry out long-term world economic analyses as a basis for
discussion of international policy issues.

In conclusion, section 4 briefly discusses the future development of the system, particularly with respect to national linkage,
which includes the integration of operational national planning models
and procedures into the UNIDO World Industry Co-operation Model system,
and the use of the system by planners and policy-makers in individual
countries.

## 2. Structure of the Model

There are two aspects of the model system which are characteristic of it. The first feature is that it is *modular* in construction: several sub-models are distinguished (see figure 1) and they are
loosely linked through information flows. Modularity allows for greater
flexibility in construction, including the incorporation of sub-models
from other sources. It also facilitates the *policy-orientation* of the
system, since, as can be seen, it allows for intervention, at all
appropriate stages, for the inclusion of strategies, goals and
scenarios.

The model is divided into two parts: the outer and the inner layer. The outer layer contains all those models, procedures, scenarios and strategies which provide the context in which the solution processes of the inner layer take place, and the components of the outer layer derive, in general, from outside UNIDO. The inner layer has been computerized within UNIDO, first in a batch version, and subsequently as an interactive programme which allows for the flexible use already mentioned.

The inner layer is a set of sub-models which analyze national economic plans and strategies in an international context. The stages of calculation are as follows: a) *National macrovectors*, where alternative macroeconomic strategies are examined; b) *international consistency*, where the national macroeconomic strategies are reconciled at an international level; c) *international trade*, where the total merchandise exports and imports of each country are broken down into their origin and destination and the bilateral flows of merchandise are then broken down into their component commodities, and d) *national input--output*, where the commodity exports of a country are examined for the effect of their impact on the individual sectors of the country.

Thus, given alternative strategies, in macroeconomic terms, for a country, the model examines these at an international level, and, through consistency analysis and trade disaggregation, derives the national sectoral implications. The principal stages within the inner layer, are shown in Figure 1 as boxes, with capitalized titles on the right-hand side of the diagram. The other boxes show the inputs and outputs of these stages. These include other exogenous assumptions with regard to economic variables and relationships, supplied either by the user of the system when operating it, or to be given by other models and hypotheses in the "outer layer", which is the left-hand side of the diagram.

The solution process of the model system will now be described, following the order of analyses outlined in Figure 1, and this is done under four headings: national macrovectors, international consistency, international trade, and national sectoral analysis.

## 2.1. National macrovectors

By this term is meant a set of alternative strategies, in macroeconomic terms, which are the starting input to the inner layer. They represent the development choices of the country concerned, based on alternative policies. They are expressed in terms of the variables shown in the box "National Macrovector Alternatives", i.e., GDP, Manufacturing Value Added, etc. How are these macrovectors obtained?

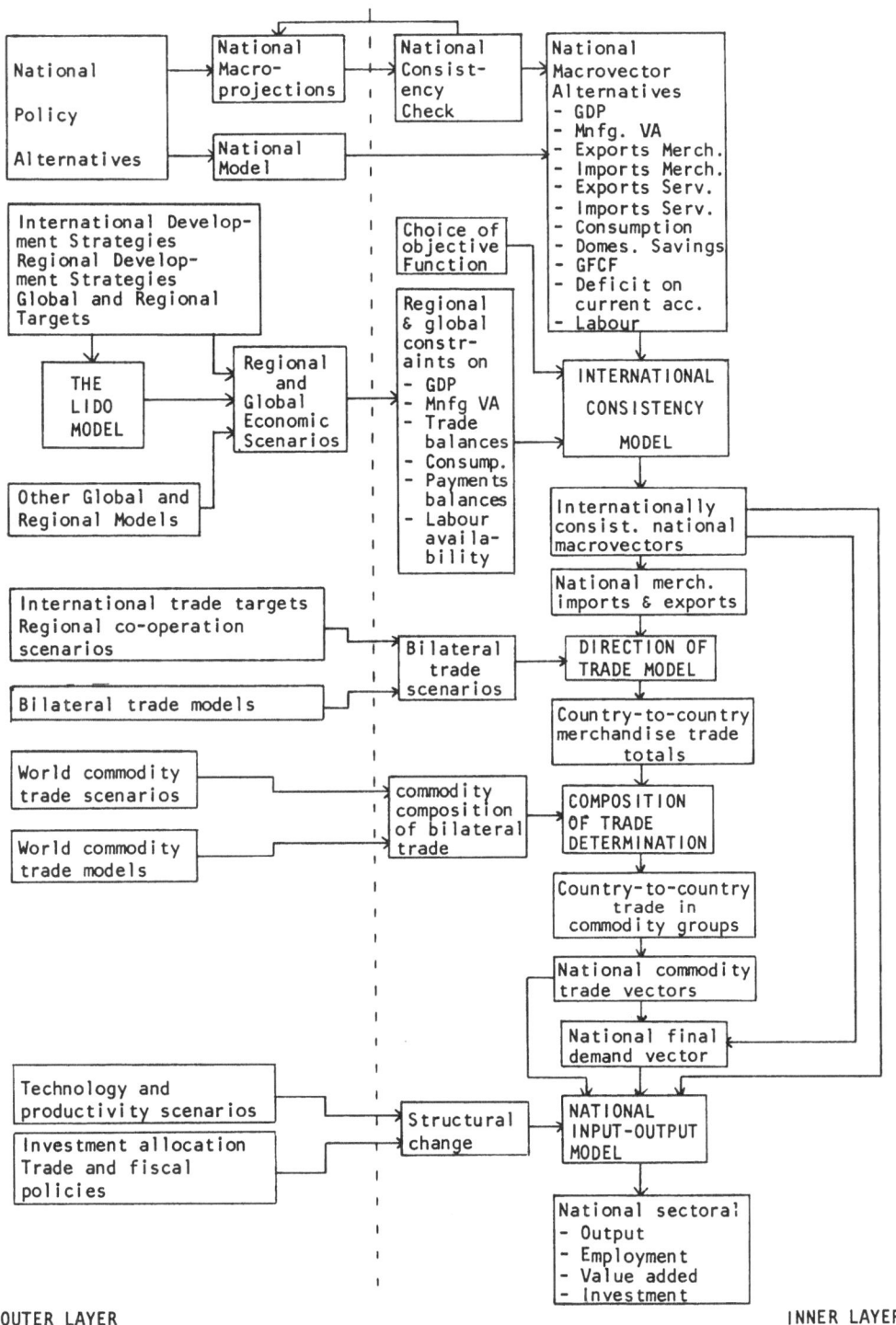

OUTER LAYER                                                                    INNER LAYER

Figure 1.   UNIDO World Industry Co-operation Model

There are two broad possibilities: they can be formed by projection, based on whatever assumptions are available, or they can be generated by a national model. In the former case, projections can be made based on such strategies as increased investment shares, continuation or increase of historical GDP growth rates, export concentration, import substitution, etc. In the latter case, if a model of the national economy concerned is available, it may be used to produce appropriate macrovectors, using whatever policy variables are contained in the model, provided only that the output of the model can be aggregated to the standard macrovector form. Thus the box "NATIONAL MODEL" can contain anything from a large optimising input-output based model with considerable sectoral detail to a small and simple national macro-economic model.

When no national model is available and the macrovectors are thus being prepared in a less formal way, there is a need to check their internal consistency. This is done at a fairly simple level in the interactive computer programme. When the user supplies a macro-vector, it is checked to see that the macroeconomic identity, between GDP and its components, is observed. If not, the user is invited to alter his input, or allow the programme to take up the slack in a selected macrovector variable. The user is warned if, for instance, the implied investment ratio is unrealistically large, and, since base year figures for 1975 are stored in the system, it is also possible to see and alter the implied growth rates and ICOR. Such national consistency checks cannot replace a national model for the purpose of generating macrovectors, but they do provide assistance in the input by the user of alternative strategies, especially if such strategies are tentative in nature, rather than the outcome of national planning procedures.

The result of this step is that there is a set (usually three) of macrovector alternatives for the country in question. The model system already contains such vectors, and base year figures, for 51 countries and regions, which together give a world coverage. The international consistency model, for which these serve as input, requires such a world coverage, but a single user is not obliged to supply or alter macrovectors for countries other than those in which he is interested, since the 1975 base year data, as well as a wide set of alternatives from previous runs of the system, is already available to him. At present, the year for which the macrovector alternatives are supplied is 1985. This is thus the solution year for the whole model system, although there is nothing in the inner layer itself which would not lend itself to solution for other years.

2.2. The International Consistency Model

This component is the means by which the macrovectors´ alternatives are reconciled at an international level, to provide a consistent solution. Since the macrovectors are derived at a national level,
there is an obvious need to ensure that world exports are equal to
world imports, or close to them within some acceptable limit of variation. However, this model also provides for consistency of another
kind, with hypotheses as to the values of regional or global aggregates,
derived from models and scenarios.

Figure 1 shows possible sources of such constraints on the
internationally consistent solution. The international development
strategies, adopted in the United Nations Development Decades, can
provide overall magnitudes of, for instance, the GDP of the developing
regions. Developing regions can also specify their desired growth
rates, labour supply, etc., and so provide regional aggregates with
which their component countries´ strategies can thus be made consistent. In addition, quantitative development targets, expressed in
relative share terms can also be incorporated. UNIDO´s particular
concern is the Lima target, which requires that the developing coun-
tries´ manufacturing be at least 25 per cent of the world total by the
year 2000. Where the targets and scenarios refer to a year other than
the solution year in question, appropriate interpolations may be made.

In this latter connection, the use in the outer layer of
other global and regional scenario models is of particular importance,
since, while not having the detail of individual countries, they may
be able to provide scenarios for world regions which have a dynamic or
optimizing character, or else have some special characteristic which
makes the generated scenario such that it provides an interesting context for the international consistency analysis. One such model is a
small, long-term world model called the LIDO (Lima Industrial Development Objective) Model, which has been constructed by UNIDO as part of
the outer layer of the system. It is described in section 3 of this
paper.

The International Consistency Model itself is a linear programming model, and a sketch of the tableau is shown in Figure 2. The
unknowns in the model are the weights w, where $w_{ik}$ is the weight
attached to the ith macrovector (alternative plan or projection) of
country k. The solution is given by finding the weights which lead to
an optimum value of an objective function. The consistent macrovectors
are formed by summing the macrovectors given for each country in the
proportions given by the weights. In fact, the problem is such that

Figure 2: Tableau of the International Consistency Model

| | Region 1 | | | Region m | World | | RHS | |
|---|---|---|---|---|---|---|---|---|
| | Country 1 macrovectors | Country 2 | ..... | Country n | | | | |
| | $P_{11}w_{11}+...+P_{1k_1}w_{1k_1}$ | $+P_{21}w_{21}+...+P_{2k_2}w_{2k_2}$ | $+.....$ | $+P_nw_{n1}+...+P_{nk_n}w_{nk_n}$ | Aggregate variables | = | b | Global constraints |
| | | | | | | | | Regional constraints |
| | $w_{11}+.....+w_{1k_1}$ | $w_{21}+.....+w_{2k_2}$ | ..... | $w_{n1}+.....+w_{nk_n}$ | | = | 1 | Weighting constraints |
| | | | | | | = | 1 | |
| | | | | | | = | 1 | |

Sample objective functions   i) $\sum\limits_i^n \sum\limits_j^{k_i} w_{ij}\, GDP_{ij} \longrightarrow$ max   ii) $\sum\limits_{i\in Dg} \sum\limits_j^{k_i} w_{ij}\, MVA_{ij} \longrightarrow$ max

For country i, there are $k_i$ alternative macrovectors

$P_{ij}$ is the jth macrovector for country i (j=1,...,$k_i$)

Dg is the set of developing countries

$GDP_{ij}$ and $MVA_{ij}$ (manufacturing value added) are the first and second elements of $P_{ij}$

non-integer solutions for the weights are found only for as many countries as there are binding constraints. In these cases, the macrovector
which gives the greatest contribution to the objective function can be
selected for each country, and thus an approximate integer solution is
obtained, since a solution which involves a linear combination of
macrovector strategies for a country is not acceptable. The linear
programming algorithm used is to be replaced with integer programming.

The principal purpose of casting the problem into an optimizing form is to achieve the consistent solution, and not to find a global
optimum as such. Nevertheless, an objective function must be specified
in order for the model to be solved, and interest attaches to its composition, since this will determine the solution which is reached.
The usual choice for the user is the maximization of world GDP, but
many others are possible, such as, for instance, the maximization of
the industrial production of developing countries.

The constraints have also to be specified, i.e., the numerical values required in the following:

## Global Constraints

  (i) Lower limit for world GDP
 (ii) Lower limit for developing countries share in world manufac-
      turing output
(iii) Acceptable percentage difference in world total exports and
      imports of merchandise
 (iv) Acceptable percentage difference in world total exports and
      imports of services
  (v) Surplus in current account to be non-negative

## Regional Constraints

      For each region
 (vi) Lower limit for the region's share in world manufacturing
      output
(vii) Lower limit for the region's consumption
(viii) Upper limit for the numbers employed

Just as it is possible to choose alternative objective functions in this model, it is also possible to choose other constraints
than those listed above. Again, future expansion of the macrovectors
to include other variables than those dealt with so far (including
agricultural production) will introduce further possibilities with
respect to constraints.

The introduction of scenarios and hypotheses of a very

general nature from the outer layer of the model system can have a very precise effect on the selection of a detailed set of individual country macroeconomic vectors. Constraint (ii) is the means by which the implications of achieving the Lima target can be examined. Similarly constraints of type (vi) can be used to describe scenarios where different regional distributions of such a target are specified.

## 2.3. Trade Determination Procedures

The international consistency model provides the starting points for a detailed analysis of merchandise trade. The analysis is in two stages:

(i) Direction of trade: the bilateral relationships between countries;

(ii) Composition of trade: the commodity breakdown of the bilateral total flows.

The consistent macrovectors contain, for each country, the exports and imports of goods and services. Because of a lack of basic data, it is not possible to proceed to a bilateral breakdown of trade in services: accordingly they are dealt with separately. Elements 3 and 4 of each (consistent) macrovector are selected to form a vector of world exports of merchandise (E) and imports (M). These are the basic constraints on the direction of trade model, for they form the row and column totals of the trade matrix which is to be found. Further information is necessary for a unique solution to be found: accordingly a matrix must be supplied which can be either a past observed trade pattern, or alternatively some projected or desired trade matrix.

This data goes to the direction of trade model, which finds, by means of a RAS updating technique,[1] the trade matrix which is consistent with the row and column totals given by the vectors E and M, and approximates to a supplied trade matrix. This supplied matrix may be a target matrix, obtained by the user´s altering or expanding submatrices of a historical matrix. Thus, for instance, he might expand that part of the matrix referring to trade among developing countries, and thus introduce a scenario of increasing economic co-operation among these countries as a component of the solution process. The result of this model is therefore a matrix of bilateral merchandise trade flows, consistent with the previously derived aggregates and

1) Previous versions of the model have experimented with alternative solution algorithms for the problem, such as a least-squared and a transportation problem technique.

approximating to some supplied pattern of international trade.

The next step is to break the bilateral merchandise trade down into decomponent commodities. This is performed in a simple manner, by using commodity composition coefficients, as follows:

$$e_{ijk} = c_{ijk} \, x^*_{ij}$$

where $e_{ijk}$ = exports of commodity k from country i to country j

$c_{ijk}$ = commodity composition coefficient for this transaction

$x^*_{ij}$ = total merchandise trade from country i to country j

$\sum_k c_{ijk}$ = 1 for all i,j

The coefficients thus specify a specific commodity composition for each transaction. They can be derived from historical data and modified by scenario analysis. The present implementation uses historical coefficients, but the computer programme allows for element, vector or submatrix adjustment of these throughout the three-dimensional matrix (i,j = 23, k = 7).

The points of intervention in these trade determination procedures are many: this is to allow for the examination of the effects of various hypotheses about international economic relations as seen through merchandise trade. Linkage in the outer layer with trade models examining such bilateral relationships allows for the supply of all or part of the initial trade matrix. Even if the outer layer model is generating only interregional relationships, these themselves can be used to modify the trade pattern already supplied.

## 2.4. National Input-Output Model

This is the fourth and, at present, the final stage in the analysis undertaken in the inner layer. In fact, the component can also be considered to be in the outer layer, if it means a national input-output model already existing and in use in the country concerned. The exact type of model (e.g. whether it is optimizing, dynamic, etc.) can vary: the common feature that all the models possess is that they are input-output based, because this type of accounting is necessary for the detailed industrial sectoral examination that is characteristic of this stage, in order to derive value added, investments, numbers employed, etc., in each branch.

The output of the previous stage, the commodity trade determination, was the breakdown of a matrix of total merchandise trade into commodity trade matrices. By then aggregating these row-wise, a vector of total exports by commodity for each country or quasi-region distinguished can be attained. The purpose of the input-output model is to analyze the implications for a national economy of such a vector of commodity exports and, more generally, the implications of the internationally consistent macroeconomic strategy that gave rise to it.

In order to do so, the export vector is treated as part of the final demand of the economy. Given that, in general, input-output models use an industry rather than a commodity classification, the usual first step is to convert the export vector to an industry classification. The sample model at present in the system, for Kenya, uses a simple historical coefficient matrix, based on analysis of industry and commodity exports, to reallocate the exports to the material sectors of the Kenyan economy. With regard to the services sectors, the total service exports of the country have been determined from the International Consistency Model, and this total by-passes the bilateral and commodity trade analysis to be disaggregated among the service sectors at a national level, by, in the Kenyan case, a set of historical coefficients.

This process produces a vector of industrial exports in the same classification as the national input-output model. Exports are, however, only one component of total final demand. The remaining components (e.g. consumption and gross fixed capital formation vectors) may be generated endogenously in the national input-output model. In such a case, the exports which have been derived are then the only exogenous component of final demand, and other exogenous information needed for this model's analysis would be derived from the outer layer. In other cases, for instance if, as in the case of Kenya, a simple impact analysis is being performed, the column totals of the remaining demand components can be derived from the internationally consistent macrovector for that country, and the related vectors generated by coefficients or elasticities from them. Thus Figure 1 shows alternative paths for the transmission of information from the international consistency model to the national input-output model.

Intervention from the outer layer into a national input-output model can take many forms. In most cases, however, it can be classified as either an imposed change in the structure or else the provision of the optimizing criteria of the model, or both. Models at present under consideration for this component are not of the optimizing type. Thus a simple open input-output model, for instance that used for Kenya, may be changed from the outer layer in a way that amounts to changing the coefficients of a set of simultaneous linear equations. An obvious example is in the area of technological change, where the input-output (A-matrix) coefficients may be altered to reflect trends and scenarios over the period in question, and, similarly, import coefficients may be altered to examine the systems response to import substitution policies. Other interventions are possible: an earlier (closed) model for Kenya had coefficients linking the second and third quadrants (giving the induced effects) which acted as policy parameters by varying the proportions in which, for instance, income generated was consumed by households or governments.

As has been said, the national input-output model forms the final stage of the inner layer. But the general diagram of the model system (Figure 1) does not show the many feedback possibilities: in other words, a number of loops could be drawn on it, to represent a wide variety of possible revisions of assumptions which can be made by the user at every stage of the analysis. At any point he may return to an earlier point in his calculations to adjust the inputs he has made in the light of the conclusions to which they have led. But the largest possible feedback loop has not yet been investigated, and thus the convergence properties of the system as a whole are as yet unknown. Thus the link from the input-output model back to the international consistency level, by aggregating the input-output model results to the standard national macrovector form, has not yet been carried out, although in the future it is hoped to do so.

3. The LIDO Model

The LIDO (Lima Industrial Development Objective) Model has been mentioned in section 2 of this paper as a component of the outer layer of the UNIDO World Industry Co-operation Model. As its name suggests, it has been designed as an aid to the implementation of the Lima target, and as such concentrates on the year 2000 and on the allocation of manufacturing production between different world areas. Its introduction was aimed primarily at two goals - firstly, the provision of consistent world and regional scenarios, incorporating the Lima target, to the inner layer, and secondly as a quick and simple substitute for the full system in circumstances where the detail of the latter was not required.

The model is a method of generating consistent scenarios rather than a forecasting or even a policy-making device. Hence it was designed to take assumptions about certain economic variables, and to trace their effects on other variables of interest, taking the achievement of the Lima target as given. The initial version was entirely static, so that it had nothing to say about the evolution of

the world economy towards the solution point for the year 2000. This
has now been altered so that the model is solved for each of the years
1980, 1990 and 2000, although the solutions themselves are still
static in nature.

The LIDO Model in its present form gives details on five
world regions (Africa, Asia, Latin America, Middle East, and industrialized countries. It uses an input-output accounting basis, dealing
with four sectors: Agriculture, Mining, Manufacturing and Others.
(This last section thus includes construction and services.)

For each of the five regions in the solution year, the model
produces an estimate of GDP and a four sector decomposition of value
added, consumption, investment, foreign trade and input-output relations. All variables are measured in constant 1975 prices. To generate these outputs the model requires three groups of inputs: firstly
the "economic technology" comprising base year information and structural relationships, secondly exogenous goals, usually the Lima target
and its regional breakdown; and thirdly exogenous assumptions about
each region´s population and trade balance and about developed coun-
tries´ growth of GDP over the projection period. The trade balances
can reflect the net transfer of resources to each area in the solution
year. The GDP assumption provides a main driving force of the model.
The GDP of one region (up to now the developed countries), determines
(through the "economic technology") their manufacturing value added;
this (through the Lima targets) implies the other (developing) regions´
manufacturing value added, and these (again via this "economic technology") give the overall economic growth rates of the developing
regions.

In slightly more detail, the solution method is as follows
(imagining a single time period, 1975-2000). The user provides GDP
growth rates and trade balances for each region. From the former the
model calculates the level of GDP in 2000 in each region and then determines the volumes of final demand for each of the four sectors. Firstly
imports are calculated using estimated import/GDP relationships.
Secondly exports are determined: if for any sector there is an exogenously determined trade balance, exports are merely this plus imports,
but otherwise all export flows, except for manufactures, are determined
by export/own GDP relationships. Exports of manufactures are always
determined as a residual so that the overall balance of trade constraint
is met. The third element of final demand is investment, which is
determined either in aggregate or by sector from the GDP assumptions -
e.g. by means of ICOR´s but the shares of each sector in the provision
of investment goods are fixed exogenously. The final element of demand,

consumption, is determined in aggregate by the GDP identity, and shared
out among sectors according to Engel curve relationships with GDP per
capita.

Having determined the net final demand for each sector's
output, the input-output tables (and their implied value-added coefficients) are used to calculate gross output and value-added by sector.
If any of the resulting sectoral growth rates appear implausible they
may be over-ridden at this stage, provided that the same sector is not
over-ridden simultaneously in each region. This feature allows, for
instance, for the imposition of an upper limit to growth in the agriculture sector of the developed regions. The excess supply or demand
generated by constraining output growth is accommodated by varying
imports.

An important feature of the determination of the gross output
and value-added figures is that the input-output tables are allowed to
vary through time: in the developed region the value-added coefficient
is assumed to grow linearly through time with other input coefficients
being adjusted pro rata to ensure that columns sum to unity, while in
the four developing regions the technical coefficients are assumed to
approach asymptotically those of the developed region, the rate of
adjustment determined by that at which their GDPs per capita approach
that of the latter region.

The next step of the solution is to compare the values added
in manufacturing against the Lima targets. If they match, a consistent
scenario has been found, but if not the GDP levels for four of the
five regions are adjusted and the process is recommenced. Since the
GDP assumption for one region is never changed, the sectoral breakdown
for that region is constant, but for the four other regions the model
continues to revise the GDP, the sectoral implications, and the input-
output coefficients themselves, until a solution is reached.

The solution process has been described as if the world
economy would evolve steadily from its base year (1975) to the Lima
targets in 2000. This, of course, is not so, for the first five years
of that period are now history, and economic performance over the next
decade or so has been largely determined by plans already in execution.
The solution has therefore been arranged in three stages. For the
first two periods, 1975-80 and 1980-90, the model is solved with exogenous growth rates of GDP for all regions:[1] this generates a complete
picture of 1990, and, using this as base, the model is solved in the
manner previously described, with the Lima targets and the developed

---

1) Since there is no target reconciliation and each area's GDP is exogenously given, these two solutions involve only one iteration each.

countries' growth rates as the main exogenous inputs. The exogeneous growth rates for 1975-80 are based on preliminary data and indicators, while those for 1980-90 are based on prospects for the United Nations Third Development Decade.

Work is now under way on the disaggregation of the sectoral detail in the model, in order to better meet the needs of policy-makers and UNIDO's own work. The basic LIDO Model merely says that manufacturing as a whole would need to grow at x per cent per annum, but policy makers clearly need more detail than that. Their ideal would be industry, or even product, level information, but that is obviously incompatible both with the simplicity of the LIDO model and with the integrity of the data. It was felt, however, that some further disaggregation was possible without overloading the system or significantly reducing the quality of the data, and so a fourteen-sector classification is being adopted.

A second development of the LIDO Model, now being undertaken, is in the area of bilateral trade. The present version uses a world pool of exports and imports, and this is being broken down into inter-regional trade by using "delta coefficients". These measure the intensity of a particular trade that might be expected as a result of the two traders' shares of world trade.[1]

An informal method of projecting these coefficients is being sought, which amounts to examining past trends, asking experts to predict the intensity of various trade links in future, considering known development plans, etc., with a view to establishing a first estimate of the $[\delta_{ij}]$ matrix, which is then updated to ensure that it meets the prior restrictions that the definitions require. Extensive time-series analysis of the determinants of changes in the delta coefficients is also being completed. At present the adjustment rule used in this updating is the minimization of the largest deviation over all elements between the first estimate of $\delta_{ij}$ and its adjusted value.

## 4. Conclusions

While the UNIDO World Industry Co-operation Model already contains a fair amount of detail, the concept of it is such that future development can take place in many areas. One of these is that of national detail. At present over fifty countries are distinguished. The overall obstacle to increasing the number is the preparation of basic projections for the remaining countries of the world. This is because, as explained, if the user of the model does not supply his own

---

1) See A. Nagy: "Methods of Structural Analysis and Projection of International Trade", Hungarian Academy of Sciences, Budapest, 1979.

projection, the model has to be able to offer alternatives itself. Thus it already contains a set of alternative projections for those countries and regions already distinguished. The future work in this area will concentrate upon completing the country coverage, especially with the collaboration of other international United Nations bodies who prepare such projections, and the United Nations Regional Economic Commissions. Another approach to the same problem will be to use the national plans of individual countries in order to derive appropriate projections, as well as to use national macroeconomic models where they are in operation.

With regard to the national input-output analysis there are a number of directions for improvement. One is to make a very simple model available for more countries. A comprehensive collection of input-output data has already been assembled. Another approach actively being pursued is to develop more sophisticated input-output based national models. This will be done using a standard software package, IDIOM, which can create for a country as sophisticated a model as there is data available. In order to foster the national linkage aspects of the UNIDO model, this latter work will be carried out in close collaboration with national planning bodies.

In the trade direction step there is not yet as much detail as in the previous one. At present there are 23 countries and regions, and so the results of the previous step have to be aggregated in order to carry out analyses of the origin and destination of merchandise trade. But it is hoped that in the future trade matrices will be available in much more detail.

With regard to the LIDO Model, while still remaining quite aggregated with respect to regions of the world, it will, as has been said, become more detailed in the number of sectors that it distinguishes, and initially these will be increased from four to fourteen. In its present form it does not include the centrally planned economies of Asia, both because these were not included in the original Lima target, and also because of the difficulty in getting suitable data for these countries, but this exclusion is also being remedied.

This short note has been intended to explain the essential features of the UNIDO World Industry Co-operation Model. It can be summarized by saying that it is a model system which is intended to provide a way for national policy-makers to examine the international effects of their proposed strategies for the future. Thus it is principally a system for others to use, which can be described as an analytical framework, negotiating tool, educational or instructional device. Its success depends on constructive dialogue with those for whom it is built.

CREDIBLE BASELINE ANALYSIS FOR MULTI-MODEL
PUBLIC POLICY STUDIES*

S.I. Gass and S.C. Parikh

College of Business and Management, University
of Maryland, College Park, Maryland 20742

Energy Division, Oak Ridge National Laboratory,
Oak Ridge, Tennessee 37830

## Abstract

The nature of public decision-making and resource allocation
is such that many complex interactions can best be examined and understood by quantitative analysis. Most organizations do not possess the
totality of models and needed analytical skills to perform detailed and
systematic quantitative analysis. Hence, the need for coordinated,
multi-organization studies that support public decision-making has
grown in recent years. This trend is expected not only to continue,
but to increase.

This paper describes the authors´ views on the process of
multi-model analysis based on their participation in an analytical
exercise, the ORNL/MITRE Study. One of the authors was the exercise
coordinator. During the study, the authors were concerned with the
issue of measuring and conveying credibility of the analysis. This
work led them to identify several key determinants, described in this
paper, that could be used to develop a rating of credibility.

## 1. Introduction

### 1.1. Use of Multi-model Multi-organization Analysis to Support Public Decision-making (23)

The nature of public decision-making and resource allocation
is such that many complex interactions can best be examined and understood by quantitative analysis. Most organizations do not possess the
totality of models and needed analytical skills to perform detailed

---

* Research performed under Subcontract No. 7867 with Dr. Saul I. Gass
under Union Carbide Corporation contract W-7405-eng-26 with the
U.S. Department of Energy.

and systematic quantitative analysis. Hence, the need for coordianted, multi-organization studies that support public decision-making has grown in recent years. This trend is expected not only to continue, but to increase.

Exhibit 1 is a schematic that conveys the idea or process of "using quantitative analysis in decision making." At the center of the process, there are elected or appointed decision makers. They receive inputs from many different sources, such as their constitutents, lobbyists, etc. Quantitative analysis forms one of such inputs. These inputs mold their thinking with regard to the problem at hand, aid them in developing plans, and reaching decisions. More often than not, decision makers have staff assistants who analyze and evaluate these inputs to identify implications of a particular decision and to develop recommendations.

The exhibit illustrates *the interaction between the decision makers and* a particular professional group that includes quantitative analysts, econometricians, engineers, operations researchers, and model assessors, i.e., *the modelers*. These professionals work with an information base that can include original (or raw) observations, as well as transformed data. The transformed data might be obtained through the use of models. The modelers use some of these data and produce transformations (in the form of computer printouts from models) which are also included in the information base. For example, scenarios and tabulations produced by the Energy Information Administration (6) might be viewed as being a part of the information base.

Within a decision-making framework, a primary goal of this group of professionals (modelers) is to provide input into the policy process. Such input flows to decision makers in one of two ways.

The first is through interaction with decision makers. It is accomplished by analysis of the information base, interpretations of the analysis and the development of insights. These varied forms of information are transmitted to the decision-makers through verbal discussions, executive summaries, charts, and other appropriate means. By this process, the decision-makers become better informed, and decisions are made with a better understanding of the results and consequences. Once decisions are made, they are documented and publicized. The second form of modeler input consists of numbers, statistical reports and quantitative information. This input is the mundane but essential task of the modelers and the type usually associated with model-based studies. It should be stressed, however, that the former type of input - that generated by modeler/decision-maker interaction - tends to be the more important and valuable one.

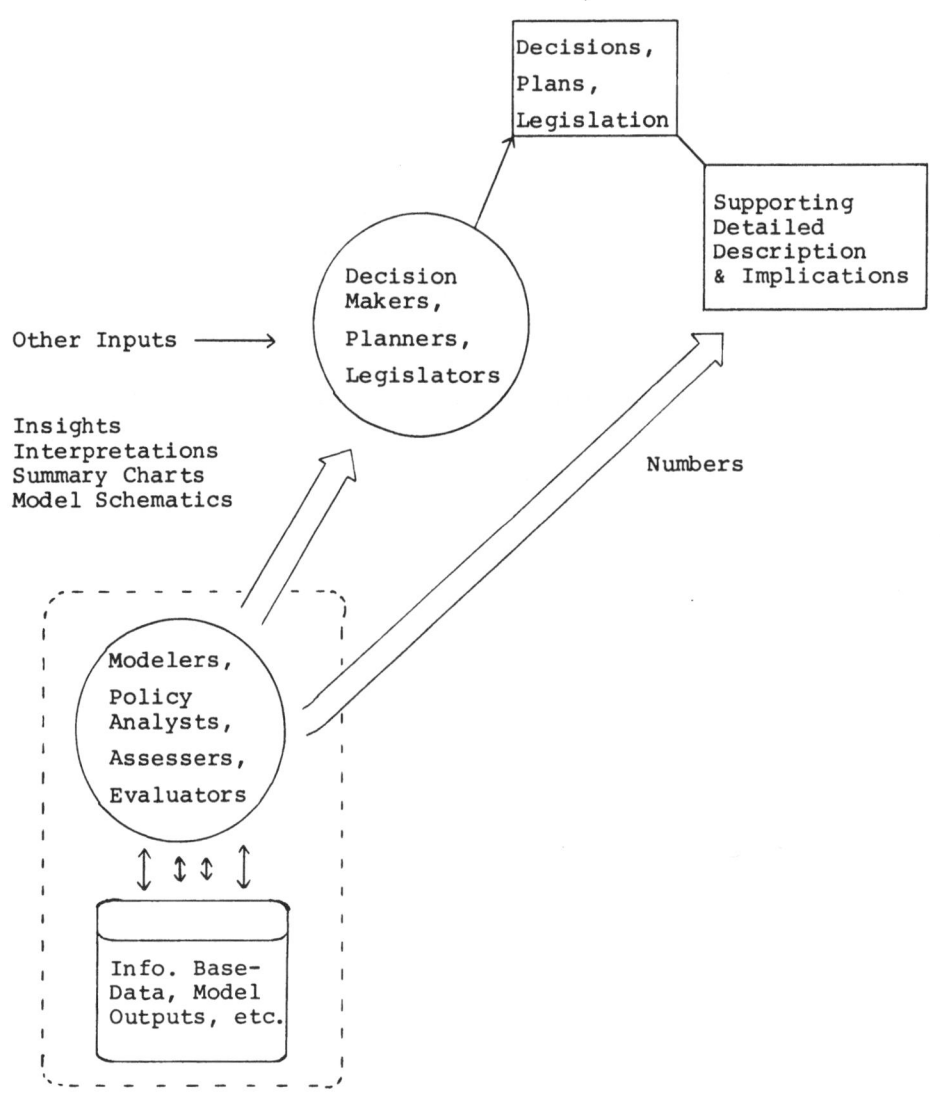

Exhibit 1.  Using quantitative analysis in decision making

Two observations are noteworthy.  First, the models and their
outputs included in the information base do not themselves produce
decisions.  The decision-makers do not directly work with information
base; the modelers and the staff personnel do.  However, implications
of many decisions may indeed be presented through numbers and tables,
thus creating an impression that the models produced the decisions.
Second, very often analysts play a dual role in that they may become
tomorrow's decision-makers.  This is another way through which

quantitative analysis can enter into the decision process. In either case, the model formulations and outputs do not produce decisions. The model information is processed by the staff and/or the decision-maker. Decisions are then reached by a human mind that somehow balances many explicit and implicit factors to manage and perform a quantitative analysis of key tradeoffs.

*The critical issue* for the profession of modelers (i.e., operations researchers, econometricians, model assessors, model developers, etc.) is to improve the quality of and confidence in the flow of the insights and in the flow of the detailed sets of numbers. Sufficiently high quality and confidence are necessary for *credible analysis*.

## 1.2. Scope of This Paper

This paper describes the authors´ views on the process of multi-model analysis based on their participation in an analytical exercise, the ORNL/MITRE Study (24). One of the authors was the exercise coordinator. During the study, we were concerned with the issue of measuring and conveying credibility of the analysis. This work led us to identify several key determinants that can be used to specify the level of credibility. The process of systematically addressing them caused us to divide the total analysis into the concepts of baseline analysis and program impact analysis. Baseline analysis forms the reference within which the program (or policy) impact analysis is performed. This paper considers the baseline analysis portion.

While the ORNL/MITRE study involved both the baseline analysis and the program impact analysis, public policy studies often involve only baseline analysis. Two prominent examples of the latter are the Global 2000 Report to the President (3) and the Energy Information Administration Annual Report to Congress (6). We thus felt it appropriate to separate and discuss the issue of credibility for the baseline analysis alone.

In section 2, we briefly describe the scope of ORNL/MITRE study to indicate how the institutional orientation of the user group may be helpful in providing a perspective on the contribution the analysis was expected to make. In section 3, we provide an analysis taxonomy that is used in section 4 to examine the determinants of credible baseline analysis. Finally, we suggest in section 5 an approach for the measurement of credibility.

## 2. The ORNL/MITRE Exercise

### 2.1. The Policy Planning and Evaluation Function of the DOE Office of Conservation and Solar Energy

The DOE Office of Conservation and Solar Energy (OCSE) is responsible for identification, assessment, and implementation of federal conservation and solar programs. This is an important function as appropriately chosen conservation and solar programs can aid in balancing the U.S. energy supply-demand requirements during the period in which the economy shifts from its heavy reliance on conventional oil and gas. For sound national energy planning, choices for conservation and solar alternative programs must be made within a strategic framework that accounts for the most likely economic and energy supply-demand futures, as well as for a range of possible but unpredictable events.

The OCSE is assisted by a staff group from the Office of Policy, Planning, and Evaluation (OPPE). In particular, OPPE assists 1) in the analysis of conservation and solar energy options, 2) the achievement of a more efficient allocation of OCSE resources to programs, and 3) the evaluation of past, present, and anticipated success of OCSE ongoing programs. The Strategic Planning and Policy Analysis Division of OPPE is responsible for 1) analyzing conservation and solar policy options, 2) evolving a strategic framework for programmatic pursuit of these options, and 3) developing analyses concerning the attractiveness of current and proposed programs. These efforts are conducted in conjunction with DOE´s annual program planning and budgeting cycle to improve the allocation of budgetary resources.

### 2.2. The ORNL/MITRE Exercise: Combining Computer Model Analyses With Policy Analyst Judgments

In December of 1979, the Strategic Planning and Policy Analysis Division initiated a major effort to unify and improve significantly the OCSE strategic planning process. This effort also included the identification and evolution of a unified approach for integrating diverse documents and projects such as multi-year programs, program impacts related to budget estimates, and legislative initiatives, etc. It called for coordinated participation of representatives of all OCSE program activities and required important contributions of OPPE contractors.

The main responsibility for quantitative analysis was assigned to two laboratories - the Oak Ridge National Laboratory (ORNL) and the Brookhaven National Laboratory (BNL). The Energy Division and Conservation Program of the ORNL were given a lead role; they were to conduct modeling and analysis activities related to estimating energy saving impacts of existing and new conservation and solar programs. Based on previous efforts of the MITRE Corporation in solar analysis, ORNL enlisted MITRE´s assistance to aid in producing a comprehensive analysis of conservation and solar options. Thus, the ORNL undertook a six-month preliminary strategy study of the estimation of energy saving impacts of current and proposed conservation and solar programs that included the participation of ORNL´s and MITRE´s computer modelers and policy analysts.

If this strategy study was to have near term relevance, its results would have to be a part of the inputs into DOE´s annual Program Planning and Budgeting System (PPBS) cycle. Therefore, it was necessary to organize the analytical effort consistent with the short time frame of the PPBS cycle. This consideration forced a tight schedule upon the analysis of the existing and proposed conservation and solar projects - a schedule that necessitated use of the current state of the art and available methodologies to perform the required analysis.

All CS energy saving impact related programs were considered in the study. There were cases for which the DOE had funded much more intensive audits of projected and reported near term energy savings. For these, the study used the DOE developed information.

Significant interaction occurred between the representatives of the program offices, OPPE staff, and ORNL/MITRE analysts. In some cases, such interaction revealed omissions or incorrect interpretations and resulted in revisions of the estimates.

The ORNL/MITRE approach consisted of estimating energy saving impacts for all program elements under a common set of energy price and economic growth assumptions. A three-step estimation procedure was followed: 1) development of the baseline projections, 2) element-by-element estimation of energy saving impacts, and 3) development of the Policy Impact Analysis Scenario (PIAS) projections that incorporate into the baseline projections the changes due to program effects.

The baseline projections provide a best estimate of the energy use in the presence of existing legislation, but without the programs. Assumptions concerning key demographic variables (such as population and labor force), macroeconomic variables (such as gross national product and disposable income), and energy variables (such as fuel prices, including the price of imported oil) formed a common set of inputs for the exercise. These data were developed by BNL and Dale Jorgenson Associates (DJA) (16). DOE´s Fiscal and Policy Guidelines provided the assumptions concerning oil import prices. Since oil needs to be imported during the study´s planning horizon, the imported oil price formed a basis for generating fuel prices for end use sectors, and for different points of end use through application of appropriate markups. Using BNL's TESOM model with these fuel prices, domestic energy production and consumption were projected. Economic growth projections were made using DJA's LITM model.

The above data were used by the ORNL/MITRE computer models for developing sectoral projections of energy use. These pooled models were the following: ORNL model of the residential sector (originally developed by Hirst (17)), ORNL model of the commercial sector (originally developed by Jackson (18)), Oak Ridge Industrial Model (being developed by Barnes, Edmonds, and Reister (25)), ORNL Highway Gasoline Demand Model (being developed by Greene and Rose (15)), Jack Faucett Associates' TEC model (9) (recently transferred to ORNL), Federal Aviation Agency's models for jet fuel computations, and MITRE Corporation's SPURR model (20) for evaluating the penetration of solar and renewable resource technologies in the buildings, industrial, and utilities sectors. The computations employed a "sequential conservation-solar interface." That is, the input assumptions were first used by the ORNL models for developing fuel and end-use projections for the residential, commercial, industrial, and transportation sectors. Next, fuel use projections of residential, commercial, and industrial sectors, together with the BNL/DJA inputs to the ORNL system, were used by the MITRE's SPURR model for developing solar use projections.

Once the baseline projections were developed, the exercise followed relatively independent and parallel assessments of energy saving impacts of the policy packages. The following DOE program offices were involved: Building and Community Systems, Industrial Programs, Transportation Programs, State and Local Programs, Energy Storage Systems, Solar Building Systems, Solar Industrial Programs, and Solar Power Systems.

The energy saving impact estimation process was not uniform across programs. For some cases, such as Buildings and Community Systems, the estimates depended largely on model-based calculations. For others, such as Energy Storage Systems, the estimates depended on ad hoc analysis. In all situations, however, it required the modeler/analyst to be experienced with the individual projects to the extent that the analyst could relate judgmentally to plausible and justifiable levels of market potential and market penetration under stated programmatic efforts.

The final estimation step consisted of identifying changes in the baseline projections due to program effects and the development of Policy Impact Analysis Scenario (PIAS) projections. The PIAS

projections are intended to provide best estimate of the future sectoral energy use in the presence of analyzed programs. Documentation
of the ORNL/MITRE study may be found in (24).

3. Baseline Analysis

3.1. The Baseline Scenario and Associate
     Baseline Analysis

In analyses such as the ORNL/MITRE study, there is a requirement to establish a set of initial conditions - baseline assumptions -
about which the analysis takes place. This requirement exists as the
problem setting must be delineated and each model cannot be operated
without fixing many input parameters that relate to the problem setting; in particular, the definition of plausible futures. The complete
statement of the baseline assumptions is essential in a multi-model
analysis. Such a statement represents the major control mechanism by
which the models and analysts are coordinated and consistency maintained
throughout the analysis. In modeling parlance, the complete statement
of the baseline assumptions concerning the problem setting and the most
plausible future is termed the *baseline scenario.*

Nearly all modeling and gaming activities involve scenarios.
According to deLeon (4), "A scenario is an account of a context or
situation created for use in a war game, a political and military exercise, or the analysis of a weapons system, strategy or problem in a
specific setting." This statement can be adapted to our situation.
The scenario represents a description of the system being analyzed.
The description is based on the model builder´s and the user´s basic
perception of the system. It also includes a clear statement of the
assumptions imposed on the model's operating environment, i.e., the
state of the model´s world. It describes the factual and/or postulated
settings of a situation and includes the objectives of the concerned
participants (2). A model-based analysis critically depends upon the
construction, definition, and interpretation of the scenario used to
generate the model´s inputs.

Although it is sometimes difficult to do, some analysts
attempt to separate the concepts of a problem´s context and baseline
scenario. The context represents the basic frame of reference and
describes the overall background or environment in which the policy
problems are considered. It sets forth the basic facts of life in the
area under study. The scenario, in turn, assumes the context and is a
description of the events leading up to a specific problem. It is the
starting point of an analysis. A scenario enables us to exclude

irrelevant material, permits concentration on the central problem under analysis, and leads to a common approach to the understanding of the subject and the future under consideration. Credibility and consistency must be ensured for a baseline scenario (1, 2, 5).

Although it represents a statement about the future, a scenario cannot be interpreted as a prediction, but only as a plausible future whose implications are to be analyzed by the modeling exercise. Scenario based analyses are conducted by varying assumptions and/or data to determine how sensitive the results are to the uncertain estimates of future conditions. Such variations usually are conducted about a reference or baseline scenario. The baseline scenario may be considered to be a description of the status quo (i.e., without policy changes) and its most plausible projection into the future.

We have used the term "most plausible" in the above discussion because of our intention to avoid specification of which of the three common measures of central tendency (mean or expected values, mode or most-likely values, and median) should be used. The choice in this regard, we feel, must be made depending upon the particulars of the application. Also relevant is the issue of treatment of uncertainty in the projections. This issue is addressed in section 3.2. below.

The baseline scenario thus provides the reference conditions that are input to the exercise models. The inputs also include the exogenous and/or uncontrollable decision elements of the problem environment; for example, assumed future GNP, population or fuel price projections. The baseline scenario enables the analysts to produce an *impact analysis* that is a result of the translation of the baseline assumptions into corresponding model inputs, running of the models, and judgmental adjustments of model outputs. The analysis can be the direct outputs of the models and/or appropriate judgmental adjustments made by the analysts. Taken together, the baseline scenario *and* the resulting impact analysis form the reference *baseline analysis*.

The need for proper specification of scenario assumptions and their model interpretations should be clear. We reinforce this need by the following examples. Most demand and supply models make some assumption concerning future population growth (22). One possibility is to assume that population grows faster and faster over time (exponential growth curve); another possibility is to assume that population levels off over time (e.g., logistics growth curve). The scenario must be explicit in this respect and all models in a coordinated exercise must be operated under a common set of assumptions.

A second possible scenario assumption might deal with a

contextual or problem environment issue, e.g., an assumption about the growth rate of oil import price. If the rate is assumed to be x%, this implies something about the level and fuel mix of energy demand. Also, the oil import price assumption must be translated into specific scenario assumptions such as numbers concerning all fuel prices.

In sum, we find that to produce credible public, scenario-based analytical studies the first order of business is the development and documentation of the baseline scenario and impact analysis that forms the associated baseline analysis.

## 3.2. Uncertainty in the Baseline Scenario

As noted, the baseline scenario represents the most plausible future. Much uncertainty surrounds many of the assumptions and data specifications of a baseline scenario. There is a need to express this uncertainty by performing standard sensitivity studies on key assumptions and/or data of the baseline scenario. This can be an expensive and time-consuming activity and, thus, it is limited usually to just a few alterations of the baseline scenario. The results of analyzing these alternate baseline scenarios represent, in a sense, the range of the possible outcomes that can occur for the assumed baseline future. Consider the following situation. For a set of baseline assumptions and data, a decision model selects a particular choice of policies that will result in a coal output over time denoted by the curve $C_b$ in Figure 1. As future oil prices are part of this baseline scenario and there is much uncertainty in these prices, two alternate baselines are used - one with higher prices and the other with lower prices than those of the baseline scenario (8). We can call them "the high price baseline scenario" and "the low price baseline scenario". These scenarios are processed by the decision models and analysts to yield the coal output curves $C_h$ and $C_\ell$, respectively, in Figure 1.

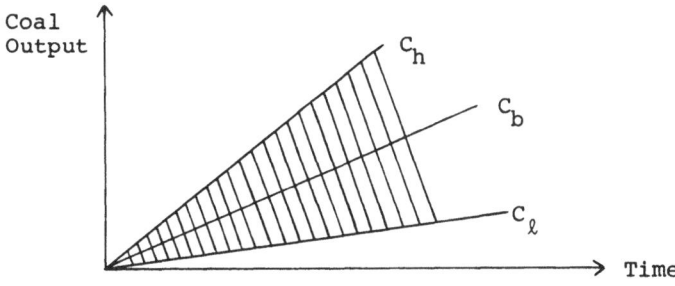

Figure 1.  Alternate Baseline Scenarios to
            Characterize Uncertainty

The shaded area in Figure 1 represents the variation in the coal output that results from the model choices as the future prices of oil are varied.

In many studies, the above type of analysis is the extent to which the models and analysts are employed. For example, the DOE Energy Information Administration (EIA) in its 1978 Annual Report to Congress (ARC) used what we would consider to be four alterations of the baseline scenario (6). These scenarios reflect high, medium and low estimates of economic growth and supply of non-OPEC energy sources. This type of "ranging" or sensitivity study of a baseline scenario serves to take into account the uncertainty in the baseline scenario.

By varying the assumptions and conditions of the baseline scenario, i.e., the views of the future, policy impacts can be analyzed over a range of possible future outcomes and compared with those of the baseline analysis. Then, depending on their assessment of the most likely course of events, the analysts and sponsors can account for the uncertainty in the exercise results. The impacts will often change with the scenario assumptions, but a consistent and controlled means of comparing policy implications of different futures is provided by a systematic use of the baseline and alternate scenarios. A comparative analysis, based on the results of a number of plausible scenarios, puts the decision-maker in a better position to decide on which policies to implement.

### 3.3. Policy Impact Analysis

In the above type of study, in which a baseline scenario is varied to reflect uncertainty, the set of choices available through the decision models is not changed. Results change only because a different choice is made under the alternate baseline scenario. Another form of analysis occurs when policy options are proposed that modifies the baseline policy specifications. The analysis then must be made using a set of choices that may (but need not) be different from the one available in the baseline analysis. Alternate scenarios are then used to produce results that can be compared to the baseline analysis to measure the impact of allowing new policy options. A case in point is the ORNL/MITRE study in which a baseline scenario is defined, along with another scenario that represents a different policy. The baseline scenario is characterized by the policy set consistent with the presence of existing legislation and no conservation and solar programs. The alternate scenario, called Program Impact Analysis Scenario (PIAS), assumes existing legislation as well as undertaking of the conservation and solar programs under current funding levels.

Policy impact analysis scenarios are subject to the same uncertainties as the baseline scenario.  To conduct a more complete policy
study, the policy impact analysis scenarios should be varied and processed in the same manner as the alternate baseline scenarios.  However,
as the running of a set of say, high, medium and low fuel price variations for each scenario is costly and time-consuming, it may become
necessary to limit the analysis to just processing and comparing the
baseline analysis and corresponding PIAS analysis.  A typical comparison
is the one in Figure 2 (from the ORNL/MITRE study) in which the level
of use of electric cars over time is shown.  The differences come about
as the ORNL/MITRE baseline and PIAS analyses are based on different
sets of policy options from which the decision logic of the models
selects the penetration of electric cars.  The uncertainty-ignoring
limitation of such analysis must be clearly understood, however.

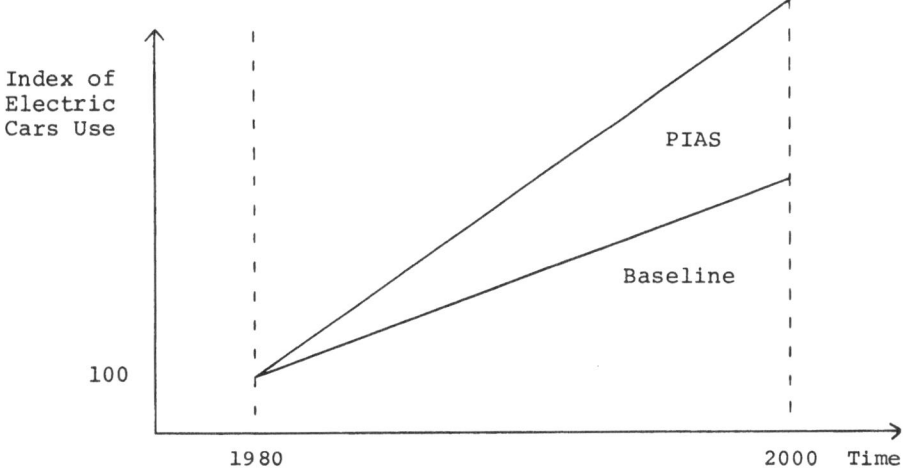

Figure 2.   Projections of Electric Cars

## 3.4. Decision Models vs Accounting Models

In our discussion so far, we tacitly assumed that the analysis
to be conducted requires *decision* (or behavioral) *models*, i.e., models
which incorporate choice-making processes to select among alternative
solutions.  Choice-making logic is incorporated in a model by simulating human behavioral actions through mathematical and logical constraints
and including a measure (possibly a utility function) for comparing and
ranking solutions.  The results of such models help form what we have
called the baseline analysis.  While the theoretical foundations and
broad validity of the behavioral representations in models cannot, in
general, be established, such models do permit systematic examination
of how the range of decisions is influenced by the baseline scenario

assumptions vis-a-vis alternate scenario assumptions. This type of comparison is the *raison d'etre* of decision models. It should be emphasized, however, that the analysis results are dependent on how the decision rules were interpreted by the model developer and how these rules were implemented in the model.

There is another class of models that is used in a policy-analysis setting as an aid in organizing and interpreting the assumptions and data of a scenario. These models, called *accounting models*, have a mathematical structure that does not incorporate behavioral assumptions. They are not used to evaluate alternative decision situations. The results of accounting models are used sometimes as inputs to decision models or other accounting models, or displayed as final results of the analysis. In either case, any transformations produced by an accounting model are merely extensions of the scenario itself and hence, they should be viewed as assumptions of, rather than a part of what we consider to be the "results" from, a policy study. This distinction needs to be made clear. Let us assume a most plausible baseline scenario, an alternative baseline scenario and a set of accounting models. If the two sets of scenario assumptions are processed only by the accounting models, then no really new decision information is produced, i.e., differences in the outputs cannot be considered to be the results of different choices among decision alternatives. Rather, the differences are due to changes in the scenario assumptions alone. The given scenarios are combined with their respective outputs from the accounting models to form extended scenarios. We cannot compare these scenarios in a decision-making sense. However, the extended scenarios can be used and processed by the decision models to aid in understanding the choice-making processes. For example, a model that projects fuel usage based on a *fixed* relationship to GNP, population, etc., is in our view an accounting model. The fuel use figures are extensions of the baseline scenario. On the other hand, if fuel use is determined by a decision model that selects among alternative use patterns based on the calculated fuel prices, then choices are made and the results add to the information base. Such additions constitute the essential difference between the baseline analysis and the baseline scenario.

## 3.5. Judgmental Adjustments of Model Outputs

As the decision logic of a model is often, at best, a primitive representation of behavioral activity, the direct output of all decision models should be scrutinized by the analysts familiar with the models. There is a need to determine if the model outputs are

consistent and acceptable within the scope of the stated scenario
assumptions. In many instances, the analysts find that the results
are unacceptable and therefore, they change the scenario, the choices
available, and/or behavioral descriptions of the model. This itera-
tive, interactive and generally ad hoc process is usually quite useful
in providing insights. However, it needs to be documented completely
with sufficient care given to how the final outputs are derived.

Although a model might have started out as a decision model,
the analyst might have restricted and changed it so that the model's
decision logic has been effectively muted and the model transformed
into an accounting model. Any added information then actually becomes
a part of the scenario. It is not to be confused with the results of
a policy study. The primary contribution of the transformed model
outputs is, then, in aiding specification of judgmental choices.

## 3.6. Other Interpretations of Baseline Analysis

As noted above, the primary use of a baseline scenario and
resulting impact analyses is to assume that the baseline represents
the most plausible future. Then, alternate baseline scenarios (possibly
two or three) are stated that attempt to capture the uncertainties of
the most plausible future. The impact analyses with reference to the
baseline scenario and the alternative baseline scenarios illustrate
the extent to which selected important uncertainties can alter the
analyses. Any differences that occur in this type of sensitivity analysis are due to the manner in which the behavioral choices of the decision models are a function of the scenario inputs. In turn, we perform
a policy impact analysis when a new scenario is hypothesized that
differs from the baseline in that the policy options of the new scenario
are (usually) different from those of the baseline. Again, sensitivity
studies can be made by varying the new scenario.

There are other possible uses of scenario analyses beyond
those described above. In some instances, we might want to define a
baseline that represents an *experimental future*, one that might represent a combination of input assumptions that is of interest but not
likely to occur. These experimental baselines can be used to investigate the logic of the model and/or produce model results that stimulate additional research or raise outcome possibilities that might bear
further investigation.

Another use of baseline analysis is to select a scenario that
represents a *desired future* or goal to be reached. The policy impact
analysis then represents assessment of the policy options that must be
implemented to reach the goal.

Finally, baseline analysis may be used to depict a *catastrophic future* that needs to be avoided. It could be used to encourage public debate and to mobilize public opinion in order to bring about policy changes. The policy impact analysis then represents assessment of the policy options that must be implemented to change the course towards a more desired future.

## 4. Determinants of a Credible Baseline Analysis

The results of a public policy study are always open to criticism - competing interests are quick to question assumptions, data and mdoel structure. They have a right to do so. Thus, it behooves the analyst team and decision maker element to be explicit and open in their activities, to document the process by which a decision was reached, and to provide the information with which the analysis can be audited and reproduced. A key part of this information is the baseline analysis, i.e., the baseline scenario and the associated model outputs, judgmental adjustments made to these outputs and the final baseline results. The exercise planners must establish, to the best of their abilities, that the baseline analysis is a credible one and can be used as a frame of reference within which the policy options are compared. We feel that the following items are the predominant determinants of a baseline analysis´ level of credibility: consistency across models, baseline scenario description, baseline analysis description, user interface, model selection, and expert assistance and review.

## 4.1. Consistency across models

The need to assure that the analysis is internally consistent is especially relevant in a multi-model, multi-analyst exercise. A basic concept here is that of *common driving variables*. These variables (inputs) provide a standard set of assumptions to the participating analysts that permit scenario runs on a common basis. A common driving variable can represent a policy option (price controls); or an assumed state of nature (oil supply); or political, economic and other world conditions (OPEC oil prices over time). Some of the common driving variables are controllable, while others are not. In either case, assumptions for common driving variables represent estimates of the future, i.e., possible states of the world, that provide a consistent basis upon which analyses are made. These assumptions are not forecasts, but represent future states the policy maker considers to be plausible when determining future policy actions.

Each common driving variable has a base case description with assumptions. The set of base case values and related assumptions form the baseline scenario.  All assumptions are input to all models.  If the common driving variable is not input to a model the analyst must then incorporate the assumptions within the modeling structure to the extent possible and note the exceptions.

Where appropriate, a common driving variable has stated variations and assumptions that are different from the baseline.  These changes correspond to common driving variable conditions that can be used to form alternate scenarios.

By defining and controlling the role of the common driving variables, the exercise coordinators can maintain "top-down" consistency.  This is quite important, for example, when a common driving variable that represents a national total must be "regionalized" by different models.  A full description of how coordination was maintained by the common driving variables must be included in the exercise documentation.

## 4.2. Baseline Scenario Description

One of the primary aims of analysis documentation is to assist outside analysts to audit the total exercise process and to verify reproducibility of the results.  We recognize that any attempt to replicate a multi-model, multi-analyst exercise may, in fact, be very difficult to accomplish.  It would probably be done rarely.  But, the adversary setting of public policy analysis requires a best effort attempt by the exercise participants to achieve a total exercise description that facilitates exercise reproducibility.  A starting point of this description is, of course, the baseline scenario.  Sufficient detail must be included that shows how the scenario addresses the objectives of and needs of the analysis.  For example, data may be available only on a national level, but the model requires regional data. How the national data were converted to regional inputs must be described.

Credibility of data, common driving variables and assumptions can be established by using accepted central sources for such information; e.g., future population data from the Bureau of Census and projections of GNP and fuel prices from cognizant DOE offices.

The credibility of the baseline scenario is also a function of the prior experience of the modelers and analysts in developing such scenarios, the qualifications of any experts used, and the process by which the insight and objectives of the decision maker (user) were incorporated.  How these resources were combined to form the details of the baseline scenario must be recorded.  In sum, the baseline scenario documentation lists the central sources of information, why they were chosen and what data are used; it describes the process by which the analysis team, experts and users influenced the choices of data and assumptions; and explains the underlying rationale for choosing and establishing the agreed upon baseline scenario (7, 10, 11, 19, 21).

4.3. Baseline Analysis Description

Total documentation of the exercise must continue into the subsequent steps. Documentation of the baseline analysis specifies how the baseline scenario was processed by the models and how the model results were used to develop the policy impact results. The decision logic of each model must be described, along with a reference to material that attempts to establish the verification of the computer program and the validity of the underlying model. In many instances such material is cursory or non-existent; but what has been done in these areas, even if nothing at all, should be stated in the baseline analysis documentation (Figure 3).

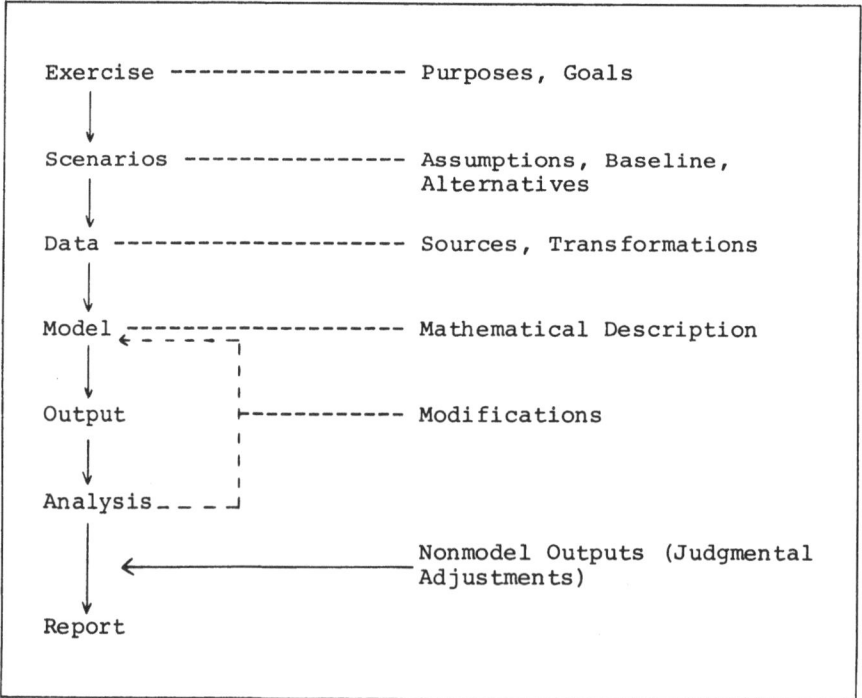

Figure 3. Documentation of Baseline Analysis

We recognize that public policy analysts cannot accept blindly the outputs of a decision model. There are always questions as to whether a model captures the decision environment properly or if the decision logic is correct. Judgmental adjustments must often be made based on the experience of the analysts, the experts and the user. But any adjustments cannot be made *in camera*. They must be described explicitly, along with the rationale for doing so.

## 4.4. User Interface

In most situations, users of the baseline analysis are the sponsor(s) of the exercise and/or the decision maker(s) who must select among the policy options being analyzed. In some situations, there might not be a user in the above sense. Many studies (and possible major exercises) are carried out by analysts/modelers who are concerned with model development, model validation and model assessment. Here, the credibility of the results are not the issue as the results will not be used in a policy-making situation. However, the more often the models are subjected to detailed study and refinements that result in demonstrated improvement and specification of the model, the more credible will be the associated baseline analysis when performed in a policy setting.

The credibility of the baseline analysis in the eyes of a specified user will suffer if the user has not been significantly involved in defining and establishing the baseline scenario and in interpreting the results of the complex study. In any event, intelligent questioning by the user will increase the overall value of the analysis and the credibility of the recommendations.

## 4.5. Model Selection

The reasons for selecting a model or a set of models to be used in a public policy study are often not stated and/or justified. A given decision situation may be so routine that a standard procedure and well accepted models may exist for its resolution. This will be the rare case. The majority of the situations will require the analyst to perform a preliminary study to determine if available models can and should be used for the new problem; if not, then plans may be made and executed to develop and test a new model. Models that are "established" and have been used with good results, especially by the current set of analysts, will tend to be adapted (if necessary) and employed in the study at hand. Past, favorable experiences will contribute positively to the determinant of credibility associated with model selection. An established model that needs no alteration, in that its

assumptions and form fit the current problem, will add a more benefi-
cial weight to this determinant than a model that has to be significantly altered and requires new computer code verification and model
validation.

An important factor to be considered is whether, for the new
analysis, the running and interpreting of the model(s) will be done by
the first-party (original) model developers. Also, any previous
second-party (original sponsor) use of the model(s) should be reported
to illustrate the "track-record" of the model(s). In some instances,
previous users of the exercise model(s) may have expressed their
measures of confidence in the model(s); if available such information
should bear on this determinant of credibility (13). Finally, third-
party (independent) assessments of the model(s) should be described
with emphasis on the how such assessments interpreted model validity
and previous uses, and the ability to use the model(s) under the
present study's assumptions (12, 14, 19).

The user of the results of an analysis study should, whenever
possible, be involved in determining which models will be employed in
the analysis. A user who has had "good" experiences with particular
models and has had a chance to measure their credibility will probably
carry over such positive impressions to the new use of the models;
faced with unfamiliar models, the user will require (rightly so) a
much more extensive set of results to develop a feeling of confidence
in the total study results.

## 4.6. Expert Assistance and Review

The credibility of an analysis study, both to the user and
the general public, will be increased if experts in the subject matter
had an in-depth involvement in all facets of the exercise. Their contributions should be attributed, along with any studies and critical
comments. We need to recognize that the use of experts can take many
forms, some of which do little to improve the credibility of the
analysis. A panel of experts that meets very occasionally to pass
judgment on the study is not of much value. We need to guard against
passive expert involvement in which a cursory review is obtained,
along with a stamp of approval. An active involvement is required.

## 5. Baseline Analysis Credibility Rating

## 5.1. Definition of Credibility Rating

Based on the total documentation of the baseline analysis,
which includes the material described in Section 4, an experienced,

independent analyst (or analyst team) should be able to translate the
information into a statement that can be interpreted as a *rating of
credibility* for the analysis.  Although we recognize that any such
rating is subjective, we feel that experienced analysts should be able
to agree, within a reasonable range of variation, as to where the
results of an analysis fall  within a five-point credibility scale.
What we propose is to use the six determinants of a credible baseline
analysis as a means of developing a rating procedure.  The six determinants are

- Consistency across models
- Baseline scenario description
- Baseline analysis description
- User interface
- Model selection
- Expert assistance and review

By reading the material that describes these six determinants, and
rating each determinant as described below, an analyst can produce a
chart that conveys the analyst's impression of the baseline analysis.
The chart contains the rating score of each determinant on a scale of
1 to 5, with 1 low and 5 high.  An illustrative result is shown in
Figure 4.

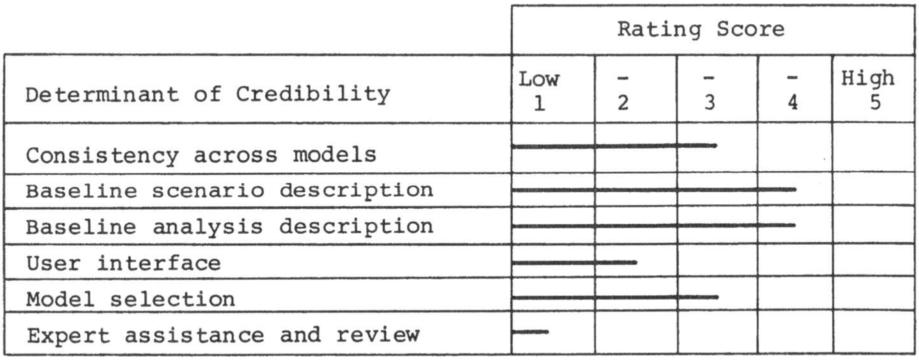

| Determinant of Credibility | Rating Score | | | | |
|---|---|---|---|---|---|
| | Low 1 | − 2 | − 3 | − 4 | High 5 |
| Consistency across models | | | | | |
| Baseline scenario description | | | | | |
| Baseline analysis description | | | | | |
| User interface | | | | | |
| Model selection | | | | | |
| Expert assistance and review | | | | | |

Figure 4. An Illustrative Credibility Rating Chart

Given such a chart, or even better, 2-3 charts developed
independently, a user can obtain a good impression of the exercise's
strong or weak points.  Further inquiries could then be initiated,
especially for determinants for which the independent raters disagree
significantly.  Weak areas would need to be strengthened for any future
studies, as well as the current one.  The raters should produce a
description that explains their rationale for the numerical score for
each determinant.

## 5.2. Implementing a Credibility Rating Analysis

The exercise coordinators should assume that the analysis product will be subjected to an examination that will result in a credibility assessment and the related critique. Thus, they must make the production of the required documentation an integral part of the study. It should not be accomplished as an afterthought or as an activity that does not have the proper resources allocated to it. The idea behind the credibility rating chart is to give the user/sponsor a much better feel for whether the study results can and should be factored into important policy decisions. The ratings also point to ways in which future analyses can be improved to be of more value to users. Hence, concern for the credibility determinants must be a part and parcel of the study.

Given the documentation of the determinants, a rater/analyst can give an overall impression for each determinant by one of five statements with each statement corresponding to the 1-5 rating scale. For example, the baseline analysis consistency determinant can be described by one of the following statements:

(Low)     1. There is no evidence of consistency.

  -       2. Consistency is maintained in a limited fashion.

  -       3. All key common driving variables are consistent in all
             models, but certain assumptions (e.g., regionalization
             schemes) are not.

  -       4. All major assumptions and key common driving variables
             are consistent in all models.

(High)    5. Complete consistency exists in all models.

Similar statements can be made about the other determinants. This leads to the joint rating chart by which the total credibility of the baseline analysis can be viewed.

## 5.3. Summary

Based on our experiences in organizing and coordinating a multi-model public policy study, we have, in this paper, expressed our views on how to obtain a credible analysis. We feel that sponsors, users, analysts and modelers can contribute much to improve our present abilities to conduct quantitative analyses by adhering to the principles, ideas, and practices that have been discussed to the maximum extent possible.

## References

1. Brewer, G. and M. Shubik, The War Game, Harvard University Press, Cambridge, Mass., 1979.

2. Brewer, G., "Scientific Gaming: The Development and Use of Free-Form Scenarios", Simulation and Games, Vol. 9, September 1978.

3. Council on Environmental Quality and the Department of State, The Global 2000 Report to the President, 1980.

4. deLeon, P., "Scenario Designs: An Overview", Simulation and Games, Vol. 6, No. 1, March 1975.

5. DeWeerd, H.A., "A Contextual Approach to Scenario Construction", Simulation and Games, Vol. 5, No. 4, December 1974.

6. Energy Information Administration, Annual Report to Congress, Department of Energy, Washington, D.C., 1978.

7. Energy Information Administration, "Model Documentation Reports" and "Analysis Documentation Reports", Memorandum, from George Lady and Susan Shaw to the DOE Applied Analysis Senior Staff, February 6, 1980.

8. Energy Modeling Forum, Coal in Transition: 1980-2000, Vol. 3, Stanford University, Stanford, CA, September 1978.

9. Jack Faucett Assoc., "TEC: Transportation Energy Conservation Model", JACKFAU-78-159-2, submitted to division of Transportation Energy Conservation, Department of Energy, August 1978, U.S. Government Printing Office, Washington, D.C.

10. Gass, Saul, I., "Computer Model Documentation: A Review and an Approach", NBS Special Publication 500-39, Washington, D.C., February 1979.

11. Gass, Saul, I., et al., "Interim Report on Model Assessment Methodology: Documentation Assessment", Operations Research Division, NBS, Washington, D.C., January 1980.

12. Gass, Saul, I. (Editor), Validation and Assessment of Energy Models, NBS Special Publication 569, U.S. Government Printing Office, Stock No. 003-003-02155-5, Washington, D.C., 1980.

13. Gass, Saul I. and Lambert Joel, "Model Confidence". Operations Research Dividion, NBS, Washington, D.C. June 1980.

14. Gass, Saul I. and B.W. Thompson, "Guidelines for Model Evaluation: An Abridged Version of the U.S. General Accounting Office Exposure Draft", Operations Research, Vol. 28, No. 2, March-April 1980.

15. Greene, D.L., and A.B. Rose, "ORNL Highway Gasoline Demand Model Documentation", Model Overview (Volume I) and User's Guide (Volume II) Energy Division, Oak Ridge National Laboratory, Oak Ridge (draft), 1980.

16. Groncki, P., et al., "Conservation and Solar Alternative Planning Case: Energy/Economy Projection", Brookhaven National Laboratory, April 3, 1980.

17. Hirst, E. and J. Carney, "The ORNL Engineering-Economic Model of Residential Energy Use", ORNL/CON-24, Oak Ridge National Laboratory, Oak Ridge, July 1978.

18. Jackson, J.R., S. Cohn, J. Cope, and W.S. Johnson, "The Commercial Demand for Energy: A Disaggregated Approach", ORNL/CON-15, Oak Ridge National Laboratory, Oak Ridge, April 1978.

19. Lady, G.M., "Model Assessment and Validation", pp. 5-22 in (24).

20. Miller, G.G., et al., "SPURR Model Capabilities", in (8).

21. National Bureau of Standards, "Guidelines for Documentation of Computer Programs and Automated Data Systems", FIPS PUB 38, Washington, D.C. February 15, 1978.

22. Oppenheim, M., Applied Models in Urban and Regional Analysis, Prentice-Hall, Inc., Englewood Cliff, N.J., 1980.

23. Parikh, S.C., "Appropriate Assessment" in (12), 1980.

24. Parikh, S.C. (Coordinator), "Energy Saving Impacts of DOE´s Conservation and Solar Programs", Oak Ridge National Laboratory, Oak Ridge, April 1981.

25. Reister, D., R. Barnes, and J. Edmonds, "Oak Ridge Industrial Model", Oak Ridge National Laboratory, Oak Ridge, July 1980 (draft).

# MEDIUM- AND LONG-TERM MODELS FOR THE ESCAP REGION

Hidde P. Smit, Rob P. Vos and Hermine J.W. Weyland

Department of Economics, Free University, Amsterdam

## 1. Introduction

Drawing up development plans and strategies should be based not only on qualitative assessments but also on quantitative analysis. Both the qualitative and the quantitative approach will only provide an adequate tool for policy recommendations if they are undertaken in a proper way. Besides, interrelationships between both ways of approaching problems of developing countries must be recognized in applying either of the two approaches. Without in-depth qualitative assessment of both assumptions underlying quantitative models and conclusions derived from modeling work, the models cannot be considered as an essential input to the formulations of development strategies.

Bearing this in mind, the Development Planning Division of the United Nations´ Economic and Social Commission for Asia and the Pacific (ESCAP) decided to revive its involvement in modeling work both for short-term and long-term projections. Short-term analysis started in late 1978 in close co-operation with UNCTAD, national institutions in ESCAP member countries and the University of Kyoto in Japan.

Long-term projection work was initiated one year earlier in late 1977 with the immediate purpose of obtaining a quantitative framework for development strategies for the United Nations´ third development decade.

Due to the lack of a team of staff members at ESCAP and because of the need for results within a year, it was decided to first consider the feasibility of using existing models and their results and conclusions. This proved not to be satisfactory. One of these modeling systems was "Future of Global Interdependence" (FUGI) developed by Professors Kaya and Onishi of Tokyo University and Soka University, respectively. They were willing to adapt the model system in collaboration with ESCAP staff members in order to better cover the ESCAP region and its development issues. The new version was called the ESCAP-FUGI model.

As is valid for any model, improvements can be achieved and its scope expanded, so that further issues can be introduced. This paper aims at describing a framework of a model which might better suit the need for quantitative analysis for the countries of the ESCAP region, giving more attention to *various types* of socio-economic structures representing ESCAP economies and particularly to the socio-politico-economic consequences of the different ways and degrees these economics are related with international trade. This framework is presented in Chapter 3. It is intended, that a study along these lines, which may turn out to be slightly over-ambitious, will be carried out as a joint project between ESCAP (the Development Planning Division), the University of Tokyo (Professor Kaya and his team) and the Free University of Amsterdam (Professor Linnemann and his team, of which the present authors are members).

In order to better assess the current status of modeling work for the ESCAP region, a brief description of the ESCAP-FUGI model will be given in Chapter 2. A more elaborate presentation can be found in Kaya, Onishi, Abe and Smit (1979) and more recently in Onishi and Kaya (1980).

2. Proposal for a Differentiated Model Framework to Evaluate the Impact of International Trade on the Socio-Economic Structure of ESCAP Countries

In a region as heterogenous as the ESCAP region a broad spectrum of development problems comes to the fore, such as starvation, poor health conditions, insufficient schooling facilities, (hidden) unemployment, high population growth, landlessness, oversized service sector, slumming of cities, low productivity, inappropriate manufacturing sector, inadequate financial resources, unequal international trade position, and so on and so forth.

Concentrating country-wise on these economic and social problems of development exploring solutions for the problems encountered, may show the following stages:

a) identifying the specific economic, social and political system and its corresponding type of development strategy, and the policies serving to stabilize and reproduce this system;

b) evaluating the existing policy framework within the rational of the system;

c) suggesting, anticipating and evaluating policies that should lead to changing economic and social conditions.

Any author, constructing a model should explicitly and does implicitly show his views on:

1. the most important development issues, the model should be or is
   focussed upon (par. 2.1);
2. the economic, social and political systems concerned and the way the
   model describes them (par. 2.2);
3. the policy analysis, i.e., the relationship between objectives,
   instruments and other variables occurring in the economic and social
   system (par. 2.3).

These three points should lead to explicit indication of:

4. economic actors and accommodating mechanisms in the "closure" of
   the economic process as a whole, which must be identical with the
   closure of the model system (par. 2.4).

## 2.1. Identification of Development Issues

Many conventional issues such as improvement of the balance
of payment, export promotion, import substitution and corresponding
employment creation and income improvements can be regarded as the core
of the economic, social and political systems and are to a large extent
explanatory to developments in living conditions of the poor. But many
of the above mentioned conventional objectives are only partial objectives as well as possible instruments to improvement in living conditions and may in turn be influenced by living conditions.

Living conditions of people may be quantified by such variables as nutrition, housing, health and education, the main basic needs
variables, with particular emphasis on such groups as the rural and the
urban poor. Quantifying such variables as calory intake per person,
access to hospital beds and square metres per person in adequate houses
for the lowest "income" groups is an extremely difficult task.

The above indicators will be dealt with on a more global level
only. Therefore three elements may be chosen, representing the core
of the development problem as well as being largely explanatory to
change in living conditions of groups in the countries concerned. These
elements are:

a. income structure;
b. income levels;
c. food production and consumption.

All three elements are directly or indirectly related with
actual international trade or its potential.

The concept income structure comprises more than income distribution. Very important aspects are distribution of assets and wealth.
Indicators such as the Gini coefficient of the distribution of personal
income pictures only a symptom of the system. Besides, it is a rather
technical measure, which is hard to interpret for dynamic systems. At

this stage, the other elements, income levels and food production need no further clarification.

## 2.2. Economic, political and social systems

The proposed project aims at modeling sets of interactions between international trade, production and income structures on the one hand and state policies on the other hand, with the objective of evaluating international trade developments in the light of income developments and the repercussions the latter will have on the former, differentiating between various politico-economic systems and the internal and external constraints these systems meet under conditions of internal and external market expansions.

It would be most desirable to study the macro structure of each country in the ESCAP region. This was attempted in the ESCAP-FUGI project, but for simplicity´s sake the same model structure was applied for each country. One should, however, allow for differences in economic and socio-political structure. In this context the following aspects come to the fore:
1) openness of the economy;
2) size of the economy;
3) population density;
4) mineral resource richness or dependence on oil imports.

Partly in correspondence with the above aspects, and partly owing to more exogenously given political factors, one has to take account of different accommodating variables per type of economy, such as types of imports, consumption, foreign or national investment, government expenditure and revenue.

In this context one needs to carefully study existing and perspective gaps and tensions invoking dynamic changes in trade, production and income structures. We think of the following gaps:
a. gaps that will need to close *ex post* as these represent financial flows:
    - import-export gap;
    - investment-saving gap;
    - public sector expenditure-revenue gap;
b. gaps in the production sphere, which may represent long-term gaps:
    - food production - basic needs gap;
    - differing productivity levels between sectors;
    - gaps between incomes of social groups.

Although, in general, the variables to be included in the model are the same for each region, causal relationships between these variables and specification of these relationships will depend upon the region concerned. The same holds true for certain aspects of the economic and social system.

It is clear that aggregation of variables in many cases reduces the feasibility to specify an equation which adequately explains the variable concerned. Many of the above mentioned indicators will suffer from too high a level of aggregation, thus need to be disaggregated. Production functions in agriculture are completely different from those in industry and from those in services. Employment creation and income generation may only be studied fruitfully on a sectoral basis. Consumption patterns will change when incomes are changing. Export in a traditional exporting economy must be explained differently from export in a region with export oriented industrialization.

Input-output analysis is an extremely useful tool to handle sectoral disaggregation. Sectoral classification should optimally be designed in such a way that resulting sectors are homogeneous with respect to
- size of farm or industry or service institution;
- traditional or modern production methods;
- tradeability.

The primary input sectors should be disaggregated optimally in such a way that a distinction can be made between social groups according to wage and non-wage income; simultaneously one may concentrate on the urban and the rural sector. The following social groups may be focussed upon
1 - self-employed, peasants, unskilled labourers;
2 - owners of small-scale enterprises, farmers, skilled workers, middle-class-salary earners;
3 - owners of large-scale enterprises and plantations, technicians, salary earners.

This distinction serves to trace the effects of change in the export, investment or consumption structure on labour and labour income and thus on income distribution. On the other hand, it must be stated that the consumption pattern is strongly related to income by social group. Investment behaviour heavily depends upon investment needs by sector and operating surplus by social group. This shows that optimally investment demand should be disaggregated by social group as well.

The above disaggregation into socio-economic groups can be made consistent within a Social Accounting Matrix (SAM). The SAM identifies the major socio-economic actors of the system according to institutions (household, corporate and state sectors), socio-economic groups and size of the firm classification. Within a SAM framework it should also be possible to trace the flows of funds between institutions and between socio-economic groups, and it should also reveal the relationships between the functional (primary) distribution of income and the "personal" (redistributed) income distribution according to socio-economic groups. Schematically disaggregation can be presented as follows:

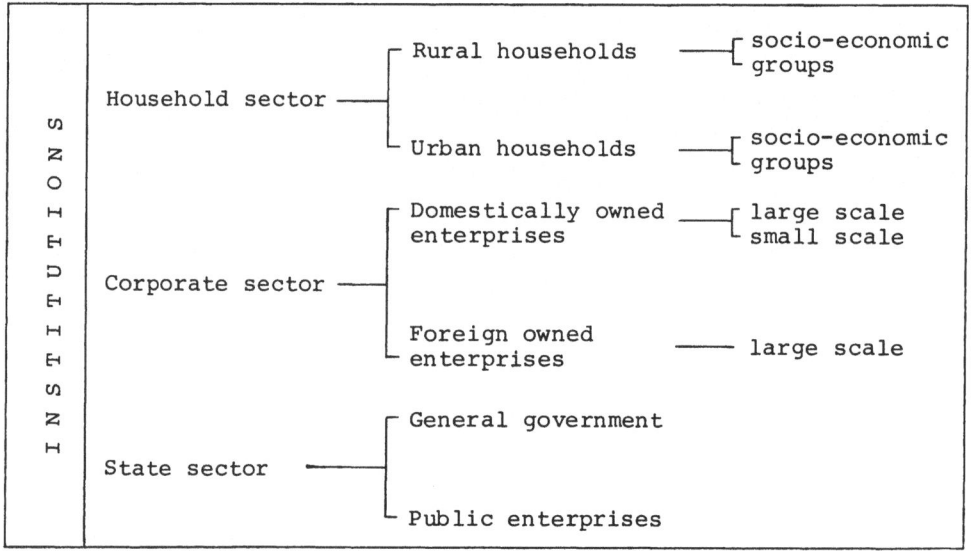

Investment ideally must be demand oriented. To transform these investments by destination to investments by origin a matrix of investment coefficients is required.

Income distribution has been introduced as the distribution of income between social groups. Within these social groups distribution seems to be fairly stable over time. This leads to the conclusion that it will not be worth the efforts of data collection and modeling to try and endogenize this kind of income distribution between workers. Endogenizing total income by social groups and size of the social groups will enable us to derive average income per worker in each social group.

Determination of the size of social groups and transfers from one social group to another can only be achieved if a demographic subsystem is included. This should include labour supply, both skilled and unskilled, in rural and urban areas. Together with labour demand by sector this will influence wages, (hidden) unemployment and under-employment and migration between rural and urban areas. Subsystems explanatory to the demographic subsystem are income distribution and basic needs satisfaction.

Nutrition, health, housing and education may be seen as basic needs. Apart from the food aspect, it is beyond the scope of this modelsystem to include detailed explanations and repercussions of (lack of) basic needs satisfaction. This will partly be covered by income levels and income structure and is partly a matter of government policy. Only food provision should play a very important role.

## 2.3. Policy Analysis

Models aiming at an evaluation of international trade relations and policies require the introduction of appropriate sets of specified instruments. These appear either as endogenous or exogenous variables, depending on the type of economy, e.g., governments can manipulate taxes, prices, wages, investments, imports, etc., but not all conceivable instruments can be used and varied in all types of economies. Or, even more complicated, in some types of economies certain variables can be influenced directly by state action, while in other types the state can only manipulate these variables indirectly. Much depends on the political economic systems of the countries concerned. But these systems are not entirely exogenously defined in themselves, but partly relate to the extent and the way corresponding economies are linked with international markets.

Tentatively, some of the relationships between the three objectives, the instruments, and other variables occurring in the economic and social system, are represented schematically in Table 2.1. The word "sector" refers to the sector of production, the word "type" refers to the type of labour or ownership. The kind of relationships between instruments (I), other variables (II) and objectives (III) may be given as follows:

I $\longrightarrow$ II    parameters and variables are affected
II $\longrightarrow$ II    "traditional" (analytic) econometric model
II $\longrightarrow$ III    qualitative and quantitative derivations of achieving
            objectives
III $\longrightarrow$ I    mainly qualitative influence on policy
III $\longrightarrow$ II    to a large extent represented in II $\longrightarrow$ II.

It must be emphasized that only the most important primary linkages are represented.

A detailed description of the model system is given in the appendix.

Table 2.1.  Links between instruments, objectives and other variables

| I – Instruments | II – Other variables | III – Objectives |
|---|---|---|
| 1 direct intervention in factor markets ⟶ II.1, II.2, II.3, II.4, II.11 | 1 production/sector ⟶ II.2, II.3, II.4, II.5, II.8, II.9, II.11, III.3 | 1 income structure |
| 2 price/wage policy ⟶ II.2, II.8, II.9, II.11, II.12 | 2 employment/sector/type ⟶ II.5, II.6, II.10, II.13, III.1, III.2 | 2 income levels |
| 3 monetary policy ⟶ II.6, II.12 | 3 investment/sector ⟶ II.1, II.8, II.11 | 3 food production and consumption |
| 4 fiscal policy ⟶ II.6, II.12 | 4 operating surplus/sector/type ⟶ II.3, II.5, II.6, II.10, III.1, III.2 | } affect most elements of I and II |
| 5 international trade arrangements ⟶ II.8, II.9 | 5 private consumption/sector/type ⟶ II.8, II.9, II.10, II.13, II.14, III.3 | |
| 6 foreign borrowing ⟶ II.7 | 6 government income ⟶ II.4, II.7, II.11, II.12, II.14, III.1, III.2 | |
| 7 private foreign investment facilities ⟶ II.3, II.8, II.9 | 7 government expenditure/sector ⟶ II.1, II.2, II.3, II.13, II.14, III.1 | |
| | 8 imports/sector ⟶ II.1, II.12, III.3 | |
| | 9 exports/sector ⟶ II.1, II.8, III.3 | |
| | 10 savings/type ⟶ II.3, II.5 | |
| | 11 wages/sector/type ⟶ II.2, II.4, II.5, II.6, II.10, III.1, III.2 | |
| | 12 prices/sector ⟶ II.3, II.4, II.8, II.9, II.10, II.11, III.2 | |
| | 13 demography ⟶ II.2, II.5, II.11, III.1, III.2 | |
| | 14 basic needs ⟶ II.1, II.13, III.1, III.2 | |
| | 15 developments in other countries ⟶ II.9 | |

## 2.4. Equilibrium, disequilibrium and the closure of the model

To gain insight into the functioning of the developing economy it is of crucial importance to identify the "closing mechanism" of the economy. In terms of formal model building this means to find the variable(s) which accommodate to an *ex post* equilibrium, when confronting means and expenditures of various socio-economic groups. The underlying adjustment mechanisms reveal the political economic functioning of the economy: consumption and savings of certain socio-economic groups are depressed in favour of the development of others. Starting from the macro-level, the *ex post* equilibrium can be shown by means of the accumulation balance, which is formulated as

$$(I_g - S_g) + (I_p - S_p) = M - E$$

The basic balancing equation represents three gaps:

- the external trade gap: $(M - E)$;
- the private sector investment-savings gap: $(I_p - S_p)$;
- the public sector investment-savings gap: $(I_g - S_g)$.

The major question to the subsequent analysis will be how *ex ante* inequality of the gaps will be reconciled *ex post* in different types of economies.

In *large, closed*[1] economies domestic adjustments will dominate by definition the process of reconciliation of the gaps (investments equalling savings). Of course, LDC economies will never be completely closed, because intermediate inputs (e.g. fertilizers) and capital goods are generally supplied largely from imports. The size of the economy will limit the scope of the external trade gap (related to GDP). Financing of investment then may happen at the expense of private consumption by means of taxation and restricted wage policies of the state. If the power of the state is insufficient to enforce such deliberate policies, the same adjustment (increase of savings) can be "forced" through monetary expansion and inflation. The foreign exchange constraint still may restrict expansion of production capacity as investments require imports of investment goods.

The *small, open*[1] economy requires, as well as may allow for, relatively more imports. Excess of imports over exports (trade gap) will mean that a domestic savings gap will be foreign financed. Investments then need not be "foreign exchange constrained". Domestic

---

1) "Large, closed" versus "small, open" economies are defined in terms of population size and the international trade quote to GDP. Open and closed thus do not refer to protectionist versus liberalized tariff structures on international trade.

adjustments can only be "delayed" to the degree that dependence on foreign capital is accepted.

More can be said on the closure of the model, when differentiating more clearly between different types of economies.

## 2.4.1. "Types" of economies

In principle, countries can be classified either on the basis of differences in the basic characteristics of the economy, such as factor supply (e.g. labour surplus, supply of natural resources) or can be based on criteria of differences in development phase or strategy (e.g. primary exporting, import substituting or industrial export specialization).

The first classification is based on geographical features (population size, natural resource endowment) and will be used for the distinction made below. This classification is being preferred as a primary distinction for the reason that differently endowed economies may have adopted equal development strategies in the course of their development process. For instance, practically all LDCs have started as primary exporters, although their endowment base need not have been suited for such a strategy. However, the term *strategy* need not necessarily mean that governments have made deliberate choices among development alternatives. The term may be used in the broader sense by which it is also related to the economy´s resource endowments. Outward looking policies based on primary exports for instance, will generally originate more in the natural resource endowments of a country (and possible historical patterns of colonialization) than in deliberate policy choices. Still, government policies may influence the "style" of the development path followed. To allow for the analysis of the role of the state, the second type of classification will be added to the distinction made on the basis of differences in factor supply.

Thus three factors form the base of the country typology:

1) The size of the economy in terms of population size and thus by the (potential) size of the domestic market. The reverse side of the factor will be the external dependence of the economy, i.e., the international trade quote to GDP.
2) Natural resource endowment in terms of mineral resources and availability of arable land.
3) Structure of land ownership, linked to population density (defined in terms of arable and accessable land to the population size) determining the institutional setting for the relationships between agriculture and industry, rural and urban labour supply and rural and

urban income distributions.

The following "types" of economies then may be distinguished:

1. Large, closed economies having a large domestic market potential due
   to their population size. External trade will be of minor importance
   in the realization of the national product.
   Two further cases can be distinguished:
   a) Densely populated country characterized by land shortage and
      labour surplus.
   b) Sparsely populated country characterized by land abundance:
      population pressure forms no major constraint on agricultural
      development.
2. Small, open economies having a small domestic market potential due
   to a relatively small population size. External trade will be of
   major significance in the realization of the national product.
   Again, two further cases may be distinguished:
   a) Densely populated country characterized by land shortage and
      labour surplus. The labour surplus condition forms a prerequisite
      for a "low wage regime", which may facilitate industrial export
      orientation as an engine to growth.
   b) Sparsely populated country characterized by land abundance. A
      significant natural resource base may enforce a tendency towards
      primary export orientation here.

Analysis will be stylized around the above "extremes", or
rather "ideal types". Of course, there exists a wide range of medium
size economies. In those cases a much closer look at specific economic
histories needs to accompany the analytical framework given below to
determine the underlying mechanisms of the adopted development strategy.
Analysis has further been aggregated to the relation between agricultural (and other primary) and non-agricultural sectors.

## 2.4.1.1. The large, closed economy

### a) Closed, densely populated economy

When considering this type of economy, one may have specific
cases in mind such as India or Bangladesh. The agricultural sector is
of outstanding importance in this type of economy and there the large
majority of the population needs to find its income source. Here the
"typical" *land shortage cum labour surplus* situation comes to the fore.
The situation has been given a theoretical framework in the classical
dualism theories, resulting from pioneering studies by W.A. Lewis
(1954). According to Lewis, the key towards economic development will
be the surplus accruing in the modern (industrial) capitalist sector.

To attain such a transition towards industrialization the process will
need to be financed through a transfer of agricultural surplus and
through low wages; the latter condition will hold as long as a labour
surplus continues to exist.

Without evaluating classical dualism theories, it can be said
that the State will be an important agent in mobilizing resources to
finance the transition. This transition towards industrialization will
have to be financed from domestic sources as reallocations of funds and
income redistribution through international trade is not a viable option to the large economy. The effort of the State to finance such a
transition may be immense under conditions of food supply shortage,
energy shortage and the lack of a domestic entrepreneurial class (low
levels of private investments). The process will be described below:

1. The institutional framework is important. Land scarcity and high
population density will lead to a great many farm plots. But this
small scale production structure may still be combined with high con-
centrations and land ownership creating rent incomes on agricultural
land[1]. The extent to which the agricultural surplus can be taxed
determines the level of the *direct* transfer of resources to the state
of modern sector investment (d $Ig \longrightarrow dT^{agr}$). Such taxation of agriculture may be termed as a form of direct *horizontal* financing. Whereas the existence of a labour surplus will depress wages to low (subsistence) levels and allows for a form of *vertical* (intra-sectoral)
financing of modern sector investment (profits versus wages).

2. Lack of coerciveness from the side of the State towards agricultural
elites may lead to the operation of other mechanisms (than taxation) to
finance a transition to development of non-agriucltural sectors. One
mechanism is changing the internal terms of trade. The wage structure
and rural-urban migration will be influenced by the internal terms of
trade, as agricultural (esp. food) prices will largely determine wage
cost. Changes in the internal terms of trade imply an *indirect* transfer of resources. The internal terms of trade will be determined by
certain conditions:

a) As in the Lewis model one may assume that a transfer of
rural labour to non-agricultural (industrial) activities will not lead

---

1) Rural elites often "suffer" from a lack of incentive to invest an
   important share of their appropriated surplus in agriculture. More
   often there is a tendency towards using the surplus for luxury consumption and/or unproductive urban investment (in real estate a.o.).
   The same hypothesis ("urban bias") is used and made plausible in
   Lipton (1977).

to a decline in productivity of food production. Rural-urban migration
of labour then will lead to a food surplus in rural areas. It seems
plausible, however, to expect that the surplus will not be completely
transferred to urban areas. Low living standards in agriculture may
lead to additional rural consumption of (part of) the extra food surplus. Higher (expected) living standards of modern sector workers may
create a situation of "excessive" urban migration, adding to the food
supply shortage in urban sectors. These factors may have a two-sided
effect: (i) there will be a tendency of increasing food prices and thus
decreasing real wages; (ii) effective demand may decline as a consequence of declining real wages and low absorptive capacity of industrial
sectors. Extra agricultural incomes may be "wasted" for reasons mentioned above ("urban bias").

b) The State may influence internal terms of trade by sup-
pressing food prices. One way of controlling food prices is to allow
for food imports. In this type of economy the effect is considered
insignificant (due to its size), unless *aid* can be acquired. Subsidizing food prices will require large state funds. The possibility of
simply fixing food prices will depend on the balance of power between
the State, rural elites and the peasant sector. However, as the bulk
of the population will earn its income from agriculture, the major
effect of a successful depression of food prices will be a decline in
agricultural incomes and thus a decline in effective demand. Absolute
poverty may thus form the constraint on economic growth.

3. If mobilization of resources through taxation or changes in the internal terms of trade prove to be insufficient (for financing d $I_g$),
the State may need to turn to deficit spending, i.e., monetary expansion as an instrument to finance investment. This process of forced
saving can only support economic growth and avoid heavy inflationary
pressures if significant multiplier effects are induced. However,
restricting ourselves to public investments, it is noted that generally
public investments, especially in early phases of development, will be
directed to activities characterized by indivisibilities and low turn-
over of social overhead capital. Then inflationary pressures will be
hard to avoid if money creation is not accompanied by increased voluntary savings. Moreover, such inflationary pressures will reinforce
the demand constraint mentioned above.

4. The above mentioned institutional constraints and problems in financing economic development have been described in a way that they will
lead to a narrowing of the size of the domestic market. These may be
summarized in terms of income distribution:

- rising rural-urban inequality through a transfer of agricultural surplus to urban sectors (either by taxation or a tendency of directing
agricultural surplus to urban sectors, i.e., the "urban bias") and
through suppression of food prices;
- intra-urban inequalities may increase from "excessive" rural-urban
migration and inadequate absorptive capacity of modern urban sectors;
- inflationary pressures from "forced savings" through rapid monetary
expansion and food supply constraints are assumed to affect wage
earners and traditional sector workers most, i.e., reinforcing inequality.

Domestic demand will be narrowed and domestic savings will
become dependent on only a small section of the population. The State
then may need to try and expand the market size. However, the same
institutional constraints as mentioned above will form a constraint on
financing extra public investment or total government expenditure.

Thus, for closing the accumulation balance in this type of
economy much will depend on the coercive power of the State. On the
one hand, private consumption (especially of urban and rural elites)
needs to be depressed to allow for the financing of high investment
rates (d $Ig \longrightarrow$ d $Cp$). On the other hand, the political nature and
power of the State may be inadequate to mobilize resources through increasing voluntary savings and taxation. "Forced savings" then may
limit the development of effective demand leading to low investment
opportunities and incentives (d $Ig_t = -$ d $Cp_t \longrightarrow -$ d $I_{t+1}$). Moreover,
the investment level will be limited by trade, as imports of required
intermediate and capital goods can only be relatively low. The foreign
exchange constraint may even be stronger if insufficient domestic food
supplies require food imports and energy (oil) shortages are significant.

b) Closed, sparsely populated economy

Given the above analyses we can briefly overview the differences between types 1a and 1b.

Under conditions of relatively low population pressure on
land the economy has the potential to produce a larger surplus. No
food supply constraints need to persist due to abundant availability
óf land and agricultural surpluses may be exported. The latter condition may finance a certain import level, facilitating a certain level
of industrialization. Some constraints may be hampering a smooth
transition to "balanced" industrial development:

- In spite of land abundance the *institutional framework* in
agriculture may create constraints in the food supply. Large scale
farming, unequal tenancy and production for agricultural exports (often

caused by colonial inheritance) may leave insufficient access to arable land for food crops. The financing of industrial development may thus be impeded by the internal terms of trade mechanism. Increasing food prices will accrue to landowners, who generally may be assumed to have a bias towards (luxury) consumption[1]. Food supply constraints may be aggravated by migration of rural labour to urban sectors. The same factor will reduce productivity in food production as labour-land ratios will decline[2]. This means that the initial "natural" tax base will be agricultural incomes of large scale landowners and agricultural export revenues. No low wage structure will emerge as in case 1a, but low real wages will only emerge through repressive wage policies of the State or of rural elites.

   - Constraints may arise from the *output structure*. If some industrialization will have taken place, it will have been inward-looking (import substitution). Import substitution behind high tariff walls may pass beyond the "easy" stage of substituting non-durable consumption goods which lasts as long as industrial inputs can be imported and domestic demand is sufficient. If the process of industrialization is extended to durables a supply constraint may arise due to shortage of capital goods and industrial inputs. The constraint may be either relieved by increasing intermediate imports (d $M_i$ = f(d I) = $m_i$ d I) or through state intervention in basic industries (d $I_g$) as indivisibilities and low turn-over will be a disincentive for private investors to invest in these sectors. The scope for d $M_i$ will be limited by the level of export revenues. Regarding the size of the economy a foreign exchange gap relative to GDP and investment effort cannot be high. Foreign direct private investments may, however, take care of autonomous (ex ante) financing of an existing external gap. Still, for the level of investment as such not to become a major constraint on economic growth, increase of public investments (d $I_g$) will be needed to relieve supply bottlenecks.

   The State will need the following problems in the process of financing economic development:

   1. As stated above, the scope for *vertical* financing through low wages will depend on repressive wage policies by the State. The *degree of horizontal* financing will depend in part on effective taxation of the agricultural sector. A tax reform (d $I_g \longrightarrow$ d T) may be needed to increase direct taxation of the landed oligarchy and urban

---

1) Thus institutional constraint impedes incentives from the price mechanism.

2) The proposition has been empirically tested, cf. Schultz (1955), Berry and Cline (1979).

high income groups or to increase state revenues through indirect taxation (e.g. on luxury consumption). The feasibility of such adjustment
will depend on the balance of power between the State (its degree of
relative autonomy) and the dominant economically powerful social groups
which are to be taxed. Both suggested forms of taxation will lead to
a downward adjustment of private consumption ($d\ I_g \longrightarrow - d\ C_p$). The
ultimate effect of the adjustment will depend, on the one hand, on the
degree of correspondence between the change in demand pattern (e.g.,
decline of luxury consumtpion) and the output structure (existing
bottlenecks, required imports), and, on the other hand, it will depend
on the effect of the tax reform on private savings.

2. Given the institutional framework in agriculture, food
supply constraints will lead through increased food prices to increased
concentration of surplus in the hands of large scale land owners. This
situation may call for state policies to move to restrict nominal wages.

3. LDC states play a significant role in allocating the
domestic savings to investing sections of society (incl. the State
itself). This can be done through taxation (increasing public savings),
but may prove to be inadequate, to close the accumulation balance.
Stimulating financial intermediation through monetary policy may be
another instrument in reconciling deficit and surplus saving entities
in the economy. Capital market imperfections may prove to be difficult
to overcome as these will be connected with structural economic constraints[1].

Monetary expansion (money printing) then may substitute for
inadequate fiscal and monetary policies, creating "forced savings"
through price increases. If nominal wage adjustments lay behind inflation, real wages will decline[2]. This will mean a shift in income
distribution towards profit incomes. The ultimate adjustment may be
a decline in private consumption as the propensity to consume from
wage and other low incomes will be higher than that from redistributed
profit incomes.

4. Public investments may induce a foreign exchange bottle-
neck as these may require imports of equipment. (The import component
of infrastructural investments - road construction - may generally be

---

1) Cf. for instance McKinnon (1973).

2) Trade unions then are assumed to be weak. One may adopt as well
the Kaleckian assumption that, under monopolistic market conditions
prevailing in many LDCs, enterprises adjust prices according to
cost increases. Profits thus are modelled as a stable mark-up on
variable costs, i.e., wages and raw materials per unit output.
Cf. Asimakopoulos (1977).

assumed to be low. Here increased state intervention in basic industries (e.g. steel) is included. Investment in these sectors will
require a significant level of imports of intermediates and capital
goods). Given the foreign exchange constraint and exogenously given
export revenues, intermediate imports for public investment need to be
compensated by a reduction in other imports like food and other consumer goods ($\bar{E} = \bar{M} = m_i (I_p + I_g) + m_c C$).

Alternatively, exports may be promoted by subsidies and tax
exemptions (reinforcing the domestic financing problem) or by exchange
rate devaluations. The latter policy will increase import prices,
again under the assumption that profits can be maintained through a
stable mark-up on production costs, these may be passed on to real
wages. Reduction of food imports may have the same effect as food
prices will tend to increase. The State´s access to foreign borrowing
may be an escape from such adjustments, but the created debt service
eventually requires real adjustments of the above kind.

Again the State is assumed to have an important impact on
the "closing" of the accumulation balance. The described constraints
of the above case differ from case (1a) due to differences in labour
supply and - as described above - different levels of industrialization. The adjustment mechanisms tend to be fairly comparable, however.
On the one hand, investments are trade limited (the difference is in
degree) and, on the other hand, major adjustment is to take place
through a decrease of private consumption. Under the above institutional conditions private consumption will adjust predominantly through
declining real wages (forced savings, food price increases).

### 2.4.1.2. The small, open economy

#### a) Open, densely populated economy

This type of economy will be initially characterized by a
*land shortage cum labour surplus* situation. The same initial constraints as in the large, densely populated economy (3.3.1.1.a) will
prevail. Available land will be intensively used and create a small
scale production structure in agriculture. Concentrated land ownership
may have forced surplus production in the form of land rent under
feudal systems. Agrarian surplus will form the tax potential to finance industrial development. The labour surplus condition will create
low income structures.

The important difference with case 1a is that the small,
densely populated economy will have through its size a greater bias
towards international trade. Export-led industrialization may be a

viable option on the basis of abundant labour supply and low wages, and
an inevitable option if natural resource endowment is poor.  In an
export-led industrialization strategy wage costs need to be kept low
as wages then form a cost element rather than that they are important
to domestic final demand.  This will induce repressive wage policies
by the State (labour market intervention) and/or food price subsidizing
policies.

In balancing the accumulation equation, the following constraints may emerge:

1. The emergence of a food supply constraint, depending on
whether the growth of agricultural production can step up with urban
population growth.  If not, the impact of an emerging food constraint
can be shifted through subsidizing food production or allowing for
food imports.

2. Technological and resource dependence combined with industrial expansion will lead to a proportional rapid increase of imports
with investment efforts: $d\,M = m_i\,(d\,I_p + d\,I_g)$.  Together with required
food imports, a significant foreign exchange gap may emerge in spite of
increasing export performance.  This gap has to be covered through
capital imports.  In part this can be done autonomously through direct
private foreign investments; aid and foreign borrowing making up for
the rest.

3. The foreign exchange gap allows for a domestic saving gap.
The two gaps, however, correspond as the rapid growth of imports will
be largely caused by high investment rates.  A high level of private
investments will need to be accompanied (or even made possible) by complementary public investments for the supply of social overhead capital.
To finance the target investment rate domestic resources will need to
be adjusted  correspondingly.  Taxing the rural sector will increase
rural-urban inequality, but its scope may be limited for reasons of
small scale production structure.  Moreover, tax income from this source
may be exempted as food prices are being suppressed to hold down nominal
wages.  Direct taxation of the urban sectors will probably increase
only proportionally with income growth.  Corporate taxes should not be
at cross purposes with policies with incentives towards investment
(tax holidays, free trade zones, etc.).  As wages already are being
kept low, these will not provide a large direct tax base either.  The
State then needs to turn to domestic borrowing, money creation (forced
savings) and foreign borrowing.

This process of restructuring capital towards export oriented
industrial sectors is accommodated in two ways: first, a high level of
imports requiring foreign financing and second, depressing private

consumption through low wages and possibly indirectly through financial
intermediation (stimulating household savings). In the long run,
foreign financing may prove to be difficult to obtain or the debt burden becomes too heavy. Therefore the trade gap will have to be closed
in time. Consequently, private consumption becomes the adjustment factor, in order to increase private and public savings to meet the
accumulation balance requirements.

b) <u>Open, sparsely populated economy</u>

One may distinguish here between natural resource rich and
natural resource poor economies. In the latter case no significant
export base exists and one would be inclined to speak rather of a small,
empty, sparsely populated country. Here, we will restrict analysis to
small, sparsely populated economies having a significant export base
(in agricultural and mineral exports).

Initially this is a typical primary export economy, comparable to case 1b. Moreover, heavy reliance on the export sector can
often be ascribed to historical patterns (colonialism), which will have
caused dependence on one or a few export products and the prerequisites
of the institutional setting: large scale farming (plantations) and
foreign ownership (esp. in mining sectors). The size of the economy
(domestic demand) will not allow for large scale import substituting
industrial development. Industrial production for domestic demand will
thus either have to be on a very small scale base, or will have to
produce for exportation as well.

The following constraint regarding financing economic development should be mentioned.

1. Domestic food supply may be or probably will prove to be
insufficient if a process of urbanization and industrial development
is induced. The cause will lie firstly in the primary export bias
and/or the extensive use of available land for this aim.

Food imports covering domestic deficits may need to compete
with high intermediate import requirements of a diversifying (industrial) investment program or may compete with imports of luxury consumer goods to satisfy high levels of conspicuous consumption by the
elites or the state bureaucracy (!) (the rate of import growth then
may exceed export growth).

2. Diversification programs, if implemented, will at first
sight have to be financed horizontally by taxing the primary export
sector. Again the power relation between the State and the dominant
socio-economic groups (the export sectors/foreign capital) will determine the extent of the transfer and the degree in which a capital outflow

from foreign dominated sectors not interested in developing production
for the (small) domestic market can be prevented.

3. Sectoral and functional income inequalities will be determined by the productivity gap between large scale production (mining
and plantations) and (most likely small scale) food crop agricultural
production. Real wages in the large scale sector need not necessarily
be kept low as productivity can be expected to be high due to modern
technologies. Wages in the modern urban secondary and tertiary and
government sector will be determined by wage structure in the modern
primary sector. Some intrasectoral (vertical) redistribution in favour
of wage incomes of a labour aristocracy may thus be enforced and may
then partially explain increasing imports of consumer goods (a possible
tendency mentioned under point 1) and a transfer of export surplus to
private consumers.

4. As far as the export sector can be expanded and some operating surplus of the export sector can be siphoned off, no financing
constraint needs to exist. Considering the domestic market constraint,
a diversification of economic activity (read industrial development)
will need to be directed towards either small scale production for the
domestic market or production for industrial exports. The former option will require protectionist policies safeguarding domestic industries from international competition. Furthermore, government policies
should be directed at diminishing cultural dependence: $(-d M \longrightarrow$
$\Delta m_c d C_p)$, in order to lower import dependence of consumer durables.
Under dominance of foreign capital groups and modern sector high-income
groups, such policies may not prove to be politically feasible. Export
oriented industrialization may be stimulated alternatively. The primary
production base and high productivity levels in large scale activities
allowing real wage increases will lead to export diversification from
the primary sector resource base as (export substitution) the most
viable option[1]. If the type of export promotion proves to be viable,
increased investment will induce a rapid growth of imports $(d M \longrightarrow$
$m_i d I)$. Required foreign financing may be found automatically if
direct foreign investments dominate capital growth. Otherwise, the

---

1) At least as a first stage of inducing manufactured export production.
A restructuring of capital towards "foot-loose" labour-intensive consumer goods production for export or assembly activities and value
added processing will find a constraint in the real wage level. This
constraint can be overcome to a certain extent through food subsidies
and/or food imports. As far as relative high real wages are allowed
by high productivity levels (high capital intensity) and labour relations in large scale sectors,"foot-loose" manufactures export production will require direct state intervention on the market factor
depressing the real wage rate.

State will need to play a significant role in transferring the export surplus to domestic entrepreneurs and itself (d Ig) through taxation and financial intermediation.

Differentiation between types of economies, phases of development and economic strategies will have an impact on:

- sectoral disaggregation and disaggregation into socio-economic groups;
- behavioural equations, the closing mechanism and the identification of the role of the State;
- the selection of the policy instruments used.

The latter is no mere "voluntaristic" issue, depending on the political will of governments in power, but will depend firstly on economic structure, and secondly on the economic and social balance of power between the State and society.

## 2.5. The composing parts of the model system

Regionalization - The model proposed here may be seen as a system of multicountry dynamic input-output models with special reference to developing ESCAP member countries. Since relationships to developed ESCAP member countries and to other parts of the world are relevant aspects regarding the future of ESCAP´s countries, these countries and regions must be somehow inserted in the system as well. Asian centrally planned economies (China, Vietnam, Laos, Kampuchea, North Korea, Mongolia, Afghanistan) are not treated on a country level but are lumped together. These considerations result in the classification by region as presented in Table 2.2.

Table 2.2.  Classification by region.

| Region number | Countries included |
|---|---|
| 1 | Korea |
| 2 | Taiwan |
| 3 | Hong Kong, Macao |
| 4 | Singapore |
| 5 | Philippines |
| 6 | Malaysia |
| 7 | Thailand |
| 8 | Indonesia |
| 9 | Sri Lanka |
| 10 | Bangladesh |
| 11 | Pakistan |
| 12 | India |
| 13 | Other developing ESCAP countries (Nepal, Bhutan, |

| | |
|---|---|
| | Pacific Islands), market economies |
| 14 | Japan |
| 15 | Australia, New Zealand |
| 16 | North America |
| 17 | European Market Economies, South Africa, Israel |
| 18 | Middle East, including Iran |
| 19 | European CPE countries |
| 20 | Latin America, Africa |
| 21 | Asian Centrally Planned Economies |

The main objective of the model system is to provide a tool for the analysis of alternative policies for developing ESCAP market economies: regions 1 to 13. The proposed structure of the model is focussed on this need. For the other regions, developed countries, developing countries outside the ESCAP region, and centrally planned economies, certain elements of the proposed model structure may be left out, since for these regions emphasis is on international trade and capital flow relationships with developing ESCAP market economies. In this chapter we will use the concept "region" meaning a country or a group of countries (cf. Table 2.2.).

Diagrammatic representation - Before discussing the composing parts of the model, a diagrammatic presentation of the broad outline of the model is given in Figure 3.1., where the model is pictured from a slightly different angle compared to the representation in Table 3.1. In order not to overload the diagram with arrows, only the most important relationships have been included.

The diagram consists of a few blocks of elements. The top layer concerns the objectives: income structure and income level (21) and gross output of food (part of 1), while at the bottom the instruments are presented: taxes (7 and 25) and other government influence (24). The elements in between show the economic and social variables, taking care of relationships between instruments and objectives and being in turn influenced by the objectives. These variables can be divided into four groups: the input variables in the first column: elements (2) to (7); the final demand variables in the second column: elements (8) to (13) and at the right the interactions and gaps: elements (14) to (20) and the input-output structure: elements (22) and (23).

Sectoral calssification of the input-output analysis for the intermediate sectors - The core of the model at a country (regional) level, linking many of the other parts of the model, is the input-output table. The sectoral calssification of the intermediate sectors must

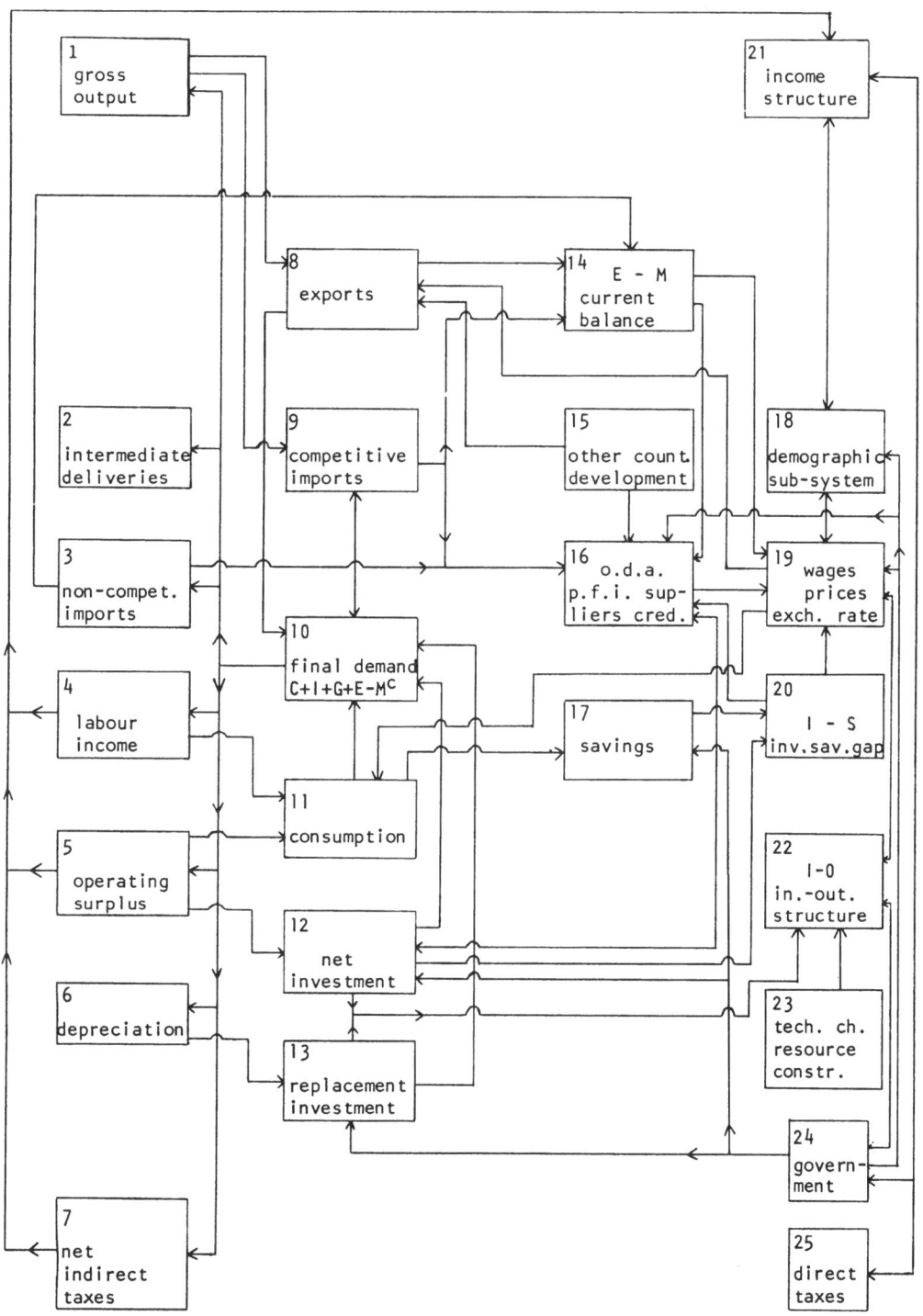

Figure 2.1. Diagrammatic presentation of the broad outline of the model.

be chosen very carefully. The sectors must be, to a large extent, homogeneous with respect to:

- size of farm or industry or service institution;
- traditional or modern production methods;
- tradability;
- formal or informal.

Many of these properties coincide, reducing the number of sectors. Small scale farming normally follows traditional production methods and has a low level of tradability for the good produced because a major part of it is produced for subsistence purposes. Modern manufacturing industries are normally set up on a large scale while the products have a high level of tradability. The informal service sector consists of small size units, not interested in tradability. These considerations lead us to the sector classification as listed in Table 2.3. The size of the production unit has in general been mentioned.

Sectors 1-5 are the primary industries. There are 10 sectors for secondary industries (sectors 6-15). The last 5 sectors cover tertiary activities. The agricultural sector is divided into 3 sub-sectors:

1 - small farming, merely subsistence;
2 - medium size farming, partly food, partly non-food;
3 - large-scale farming, mainly non-food.

Sectors 2 and 3 may be export sectors. The mining sector has been split up to distinguish between energy and others. The disaggregation of the manufacturing sectors is based on tradables (6-11) and non-tradables (12-15). The order in which sectors for tradables are listed broadly reflect an increasing level of industrial development. Non-tradable sectors have large infrastructural contents. In the service sectors, informal services have been separated because this is where the (un(der)) employed try to "earn" a living.

For some countries it may be useless to follow such a rather high level of disaggregation. Some industrial sectors may be lumped together. On the other hand, it is possible to introduce other or additional criteria for sectoral disaggregation. One may think of more emphasis on separation of production for the domestic and for the export market and/or of production for consumer goods or industrial inputs. As it is intended, that international trade begin at the sectoral level, it is necessary to apply a largely similar sectoral disaggregation for all countries to be included.

Tabel 2.3. Sectoral classification of the input-output tables.

1     agriculture and fishery (small)
2     agricultural farms (medium)
3     agriculture and forestry - plantations (large)
4     mining excluding coal and petroleum (large)
5     petroleum, coal, etc. (large)
6     food, beverages, tobacco, textile, wearing apparel, leather,
      wood products (small)
7     ditto (large)
8     chemical products including petroleum products (large)
9     metal and products and non-electrical machinery (large)
10    electrical machinery (large)
11    transport equipment (large)
12    construction (small)
13    construction (large)
14    electricity, gas (large)
15    water, including watercontrol (large)
16    merchandise, trade and commerce (large), including banking
      and communication
17    merchandise, trade and commerce (small)
18    services (large), government and private, including
      education, housing, health
19    services, formal n.e.s.
20    services, informal.

Disaggregation of value added - Not considering net indirect
taxes, which will be discussed at a later stage, value added may be
divided into labour income and operating surplus income. Since this is
the basis for the determination of the income distribution, which in
turn affects consumption investment and savings patterns, it is essential to do a very careful job in modeling the value added component of
the countries' economies, although this may be difficult, particularly
for small farmers and service sectors.

     To start off, it is extremely useful to distinguish between
rural and urban areas because of differences in production structure,
consumption structure, wage levels, aspects of un(der)employment, etc.
Further, differences in levels of skills are important factors in
production functions, wage determination, employment opportunities,
etc. Therefore, it is necessary to distinguish between unskilled,
skilled and highly-skilled labour. Finally, ownership scale, as one
of the determinants for operating surplus, is relevant to production
functions, consumption and investment behaviour, tradability and foreign

influence. Thus we arrive at a rather optimal division of population
into social groups as depicted in Table 2.4.

Table 2.4. Population by social groups

Population and ownership in the rural areas

    1a.  peasants, small farmers
     b.  landless labourers, unskilled
    2a.  farmers
     b.  skilled workers
     c. large-scale farmers

Ditto in the urban areas

    1a.  self-employed (3 employees)
     b.  unskilled workers
    2a.  owners of small-scale enterprises (10)
     b.  skilled workers (blue collar)
     c.  middle-class - salary earners (white collar)
    3a.  owners of large-scale enterprises
     b.  technicians
     c. salary earners.

        Categories b and c always indicate labour whereas the first
(a) group represents entrepreneurs/owners. In this aspect, special
attention must be paid to the role of the State. Most likely categories 1b. and 2b. in the rural areas will be aggregated due to data
availability (ILO workers´ classification). Important differences
between classes of ownership are access to investment goods and types
of labour employed. In general, both for rural and urban, 1a. has
low access to investment goods, 2a. has a medium level and 3a. has a
high level of access to investment goods. The relationship between
ownership/farm/firm and types of labour is schematically represented
in Figure 2.2.

| Skilled workers | Ownership scale | rural | | | urban | | |
|---|---|---|---|---|---|---|---|
| | | 1 | 2 | 3 | 1 | 2 | 3 |
| 1 unskilled | | x | x | x | x | x | x |
| 2 skilled, salary earner | | | | x | | x | x |
| 3 technician professional | | | | | | | x |

Figure 2.2. Employment of labour by type by farms/firms by size.

Demographic subsystem - Before describing the production
system and explaining such variables as investment and labour demand,
we should consider labour supply, both skilled and unskilled, being
constraints to as well as partly explanatory to demand for labour and
investment goods.  Discrepancies between demand and supply of labour
will lead to:
- unemployment/underemployment
- changes in wages
- migration
- changes in the reservoir of the informal service sector.

A schematic representation of the demographic subsystem is
given in Figure 2.3.  Subsystems explanatory to the demographic subsystem are income distribution and basic needs on the one hand and wages
and prices on the other.

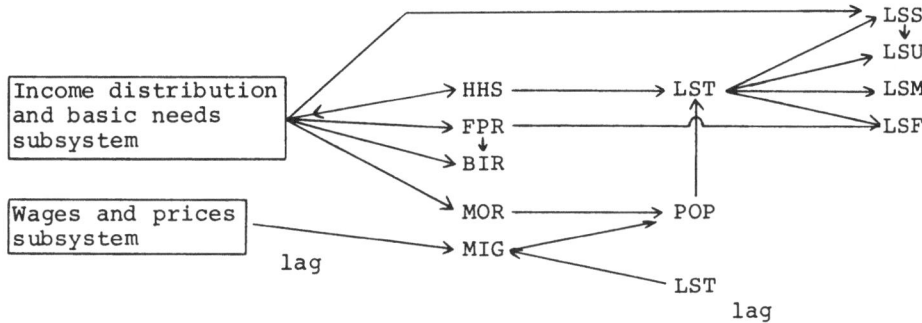

Figure 2.3.  The demographic subsbstem

where BIR = birth rate
      FPR = female participation rate
      HHS = household size
      LSF = labour supply, female
      LSM = labour supply, male
      LSU = labour supply, unskilled
      LSS = labour supply, skilled
      LST = labour supply, total
      MIG = migration rural-urban or vice versa
      MOR = mortality rate
      POP = population.

This subsystem is valid both for the urban and the rural area.
Adding an R or a U to each of the variables gives the rural and the
urban component separately.  The two may run without interdependence,
provided that MIGR = - MIGU.

Some remarks may be made:

- The indicated relations between income distribution and
household size are highly probable to run in the following direction:
the larger the household size the better the distribution parameters
look, because of pooling of incomes and informal transfer mechanisms
within the extended family.  It has also been found that low incomes
of an informal nature are more often than not associated with complementary earning functions rather than serving as head-of-the-household
incomes.

However, there is also a feed-back from income *level* to
household *size*: the lower the level, the larger the household size.
But such a relation holds only cross-country wise.

- Female participation and birthrates appear to' correlate
only in developed economies or in high-class brackets of underdeveloped
economies.  In extended family environments the relation does not hold
because of the pooling of services effect.

- Migration is supposed to be a lagged response to differences in wages (and price) levels between the urban and the rural sector, representing discrepancies between labour demand and supply.  Of
course some gap between urban and rural wages is considered normal.
Thus

$$POPU = POPU_{-1} (1 + BIRU - MORU) - MIGU.$$

A similar equation holds for POPR. MIGU = migration from urban to
rural areas.

Production functions for manufacturing industries - Thus far
the model has been embedded in the input-output framework.  This implies that, essentially, a production function must be postulated
assuming fixed wage and profits coefficients.  However, this is highly
unrealistic.  It has been found that small-scale industries have
notoriously low substitution elasticities and large-scale units may
show unexpectedly high ones.

If substitution elasticities between wage and capital are
low, depressed wages do not stimulate the hiring of labour.  This
means that value added shares for wages will decline.  As a result,
income distribution will be seriously affected.  For large-scale units
with high substitution elasticities depressed wage levels may mean
employment creation, and consequently positive distribution effects
may be expected.  The model should be capable of pointing out such
effects which may be at the root of the differences in income distribution trends observed for countries like Korea and Taiwan on the one
hand and—India and Indonesia on the other.

Presumably, it is not far from reality to accept at the sectoral level the fixed share of the aggregate of labour, operating surplus and other value added components, giving room for the various value added components to adjust simultaneously within this boundary. This allows us to formulate production functions with substitution for the relationship between production, labour, capital and other variables.

The model might be formulated as follows:

$$X_i = \sum_j \alpha_{ji} X_i + Y_i \qquad (2.1.)$$

where $X_i$ = gross output of sector i

$\alpha_{ji}$ = input-output coefficients intermediate sectors

$Y_i$ = value added of sector i.

Assuming fixed $a_{ji}$, a production function for $X_i$ is essentially the same as a production function for $Y_i$ because

$$Y_i = (1 - \sum_j \alpha_{ji}) X_i \qquad (2.2.)$$

For $Y_i$ a CES or Cobb-Douglas production function may be formulated:

$$Y_i = f (LDS_i, LDU_i, K_i, Z_i) \qquad (2.3.)$$

where $LDS_i$ = input of skilled labour

$LDU_i$ = input of unskilled labour

$K_i$ = input of capital goods

$Z_i$ = input of other production factors including variables for technical progress.

In order to better permit estimation and further analysis, it may be preferable to relate $\Delta Y_i$ to $\Delta LDS_i$, $\Delta LDU_i$, $I_i = \Delta K_i$, $\Delta Z_i$. Optimization will lead to values for $\Delta LDS_i$, etc., if gross output $X_i$ is given or estimated and thus $\Delta Y_i$ can be obtained. For each of the input categories for labour and for capital the respective input coefficient $b_{1i}^{\beta_{1i}} \cdot \tilde{b}_{1i}^{1-\beta_{1i}}$ \qquad (2.4.)

where $b_{1i}^*$ = new input coefficient

$b_{1i}$ = old input coefficient

$\tilde{b}_{1i}$ = optimal input coefficient for incremental output

$\beta_{1i}$ = weight for adjustment.

The weight $\beta_{1i}$ must be based on the share of last year´s production factors in this year´s production factors adjusted for the degree of penetration of new methods.

Replacement investment may be treated similarly. Actual

operating surplus may differ from calculated operating surplus using
the above coefficients because of wage and price developments. If
access to investment goods is constrained the above procedure must be
amended.

Production in the agricultural sector - The agricultural
sector is one of the sectors where investment decisions may be made in
a somewhat different way. Since some other aspects of the agricultural
sector are important as well in determining production, labour demand,
investment and last but not least migration and income distribution, we
shall pay some more attention to this sector.

Production in the agricultural sector is performed quite
differently by small farmers compared to larger farmers producing
either for the urban market or for the export market. The latter group
has been disaggregated into medium farmers and large farmers (+ plantations). Inputs of intermediate goods are taken care of by the input-
output table. Production should be related to several variables
depending on the scale of the farm or estate.

For the peasant or small farmer, investment in stocks and in
machinery largely depend upon income. The same is valid for transport
equipment, while constructs might be related to income and stock formation. Production then can be explained from area, adjusted upwards
for use of fertilizers and labour quality (labour quality may be represented by basic needs variables, health, etc.), investment (machinery
and equipment, etc.) and technological improvements. New addition to
land may, on the other hand, insert decreasing returns to scale,
owing to the, on average, lower quality of additional land. Finally,
area per worker will influence productivity. Labour demand equals
labour supply since it is assumed that there is no open unemployment
in the rural areas.

The larger farmer, and certainly the estate, base investment
decisions about machinery and equipment on income and on factor costs
as well. Construction investment is related to stocks and to machinery,
whereas transport equipment is rather complementary to production.
Capital plays a definite role in the production function. Labour
demand, both skilled and unskilled, can be explained from production
and investment, thus shifting excess labour supply to the group of
subsistence farmers and putting pressure on migration from rural and
urban areas. The same variables occur in the production function as
for the small farmer. We should add to this water control (irrigation,
etc.) which improves productivity. This should be seen as investment
by the government (I-0 sector 15). A schematic representation is
presented in Figure 2.4.

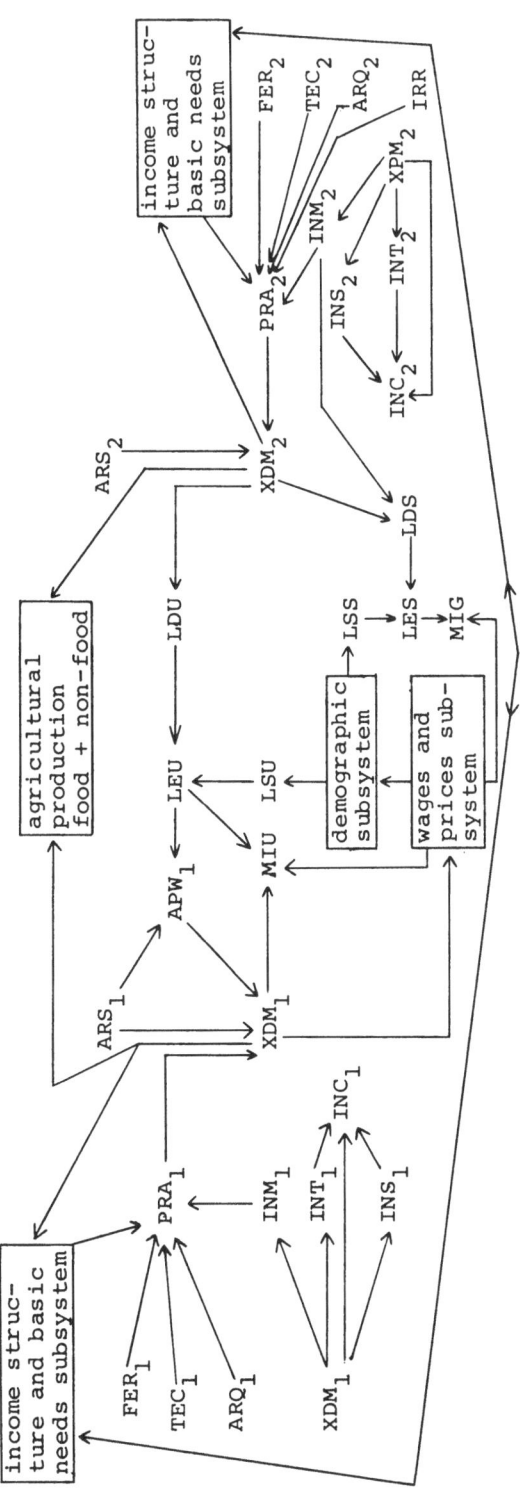

Figure 2.4. Production and employment in agriculture.

where: subscript 1 = small farmers
2 = medium and large farmers

APW = area per worker
ARQ = area quality
ARS = area size
FER = fertilizers
INC = investment in construction
INM = investment in machinery and equipment
INS = investment in stocks
INT = investment in transport equipment
IRR = irrigation
LDS = labour demand, skilled

LDU = labour demand, unskilled
LES = excess labour, skilled
LEU = excess labour, unskilled
LSS = labour supply, skilled
LSU = labour supply, unskilled
MIS = migration, skilled workers
MIU = migration, unskilled workers
PRA = productivity per hectare
TEC = disembodied technical progress
XDM = domestic production (gross)

Production functions for extractive industries - In general,
mining is large scale business, where production is supply or demand
determined. Production is highly capital intensive and access to investment goods is no reason for the substitution of capital by labour
or vice versa. Fixed input coefficients, though possibly changing
over time owing to technical progress, seem to be the best approach.

Production in the merchandise and service sectors - Because
of essential difference between the formal sectors in this group (sectors 16-19) and the informal service sector, the two groups will be
discussed separately.

Production in the formal sectors is rather demand or government policy oriented. Investment largely consists of constructs,
although, e.g., in the case of banking and communication. Machinery
other than those necessary for buildings, must be taken into account.
For all these sectors substitution between capital and labour is
rather irrelevant.

The informal sectors almost by definition do not require
capital goods. The un(der)employed and the self-employed in this sector must be treated as a residual, a reservoir in the cities, with
two ways in and out: the manufacturing and formal service sector in
the cities or the rural areas through migration. "Production" is
largely demand oriented, but the size of this sector may add pressure
to increase output.

Demand for investment goods - Demand for investment goods
must be divided into replacement investment and net investment. Afterwards investment by destination must be transformed into investment by
origin which can be fitted into final demand. We shall discuss the
various components.

Ideally, replacement investment is derived from gross output
using coefficient in value added from the input-output table. It may
be necessary, however, to adjust this procedure, relating replacement
investment to capital stock or lagged investment.

Net investment has been discussed above. It may be derived
from production increases, using the production function. If supply
of capital for investment is limited in some sectors, it may be necessary to reduce investment. If aggregate supply of capital is insufficient because of lack of funds from savings and other sources, it is
necessary to reduce overall investment. Distribution of this reduction over the sectors must take account of total investment requirement per sector and a factor representing the power of a sector to
attract limited supply of financial resources, be they domestic or
foreign.

After arriving at gross investment being replacement plus net investment by destination, a matrix has to be determined redistributing gross investment by destination to gross investment by origin, similar to the matrices used in theoretical dynamic input-output models. Thus gross investment can be included in a column for final demand in the input-output analysis. Information on the transformation matrix must be partially obtained from micro studies.

Consumption and savings - Basically, the consumption functions are quite simple. Consumption, savings and direct taxes are the composing parts of expenditure. Marginal savings quotes and direct tax ratios to income differ per social group and should therefore be dealt with in a rather disaggregated way. This allows for effects of redistribution of income on consumption and savings patterns. The model may be formulated as follows:

$$T_g = \tau_0 + \tau_1 YW_g$$
$$S_g = \sigma_0 + \sigma_1 (YW_g - T_g)$$

where $S_g$ = savings of social group g, g = 1, ..., G

$T_g$ = income taxes of social group g

$YW_g$ = total income of groups g

The last concept, total income of group g, will be presented at a later stage, as it requires labour as well as wages and operating surplus. Possibly G must be adjusted for average income levels. Now consumption expenditure can be directly related to income minus savings and taxes. In order to arrive at consumption data per sector of production, consumption functions should be set up for each sector i, i = 1, ..., n. We should allow for the possible influence of prices, resulting in consumption shifts from one sector to another.

$$C_{i,g} = \alpha_{i,g} + \beta_{i,g} (YW_g - S_g - T_g) + \lambda_{i,g} P_i$$

where $C_{i,g}$ = consumption of products of sector i by social group g

and $P_i$ = price of the product of sector i.

After deriving consumption functions per sector for each social group, total consumption of this social group $C_g$ should equal the sum of the consumption components per sector.

$$C_g = YW_g - S_g - T_g$$

It may be necessary to include prices of key sectors in the savings equation of the low-income social groups, because for example a high increase in food prices may decrease savings in urban areas.

In cases where data and relationships for depreciation are

very poor, it may be more appropriate not to distinguish between gross
and net investment and to introduce the concept of gross savings including both net savings and expenditure on replacement investment. However,
this is a rather unattractive way out.

International trade - The international trade part is one of
the most important parts of the model. It describes relationships and
sensitivity to developments in other countries and regions and as such
it makes the country concerned part of world wide developments. On the
other hand, exports and imports may be heavily influenced by the domestic scene and may be elements of policy formulation and complementation.

Focussing on exports first, it is clear that exports are
largely demand oriented. Import needs in importing countries must be
considered major causes for export levels. Aiming at the possibility
of including such issues as traditional export promotion and export
oriented industrialization, it is necessary to analyze exports at the
sectoral level. Besides, as one importing region grows faster than
others and on the other side the issue of growing trade between developing countries comes to the fore, it is very important to focus on
bilateral trade flows as well. If data or computational limitations
prohibit analyzing bilateral trade flows at the sectoral level, it may
be possible to analyze bilateral trade on broad sectoral levels and to
design a device for disaggregation at a later stage. In such a case
one might come up with the following classification by broad sectors
(tradables only).

    a - agriculture (sectors 1-3)

    b - mining (sectors 4-5)

    c - manufacturing (sectors 6-7)

    d - chemical industries (sector 8)

    e - other manufacturing (sectors 9-11)

    f - merchandise and transport (sectors 16-18)

    g - services (sector 19).

The device for disaggregation must contain national, international and sectoral consistency requirements. For some sectors it
may be necessary to explain exports as supply oriented or supply constrained. Certain constraints are necessary for any sector, because
of slow changes in manufacturing structure.

On the import side, it is fashionable to distinguish between
competitive imports and complementary (non-competitive) imports. A
problem occurs in defining whether a product (output of a sector) is
competitive or not and if it is competitive whether importing that particular product results from insufficient capacity or from inappropriate
capacity, which is more a qualitative criterium.

The role of these three elements of imports is rather different.  Complementary imports may be related directly to domestic production, e.g., rubber imports into Korea.  Competitive imports may emerge
from domestic supply insufficiencies (shortage of rice) or, for example,
from inadequate technology (imports of cars).  Treatment of these two
parts of competitive imports must be different; where the first part
is introduced to bridge the gap between demand and supply at the sectoral level, the second part may be related to such factors as income,
balance of payments position and industrial development in the country
concerned.

Let us first describe an example of the kind of analysis useful for the model.  A hypothetical situation for three sectors is
presented in Table 4.4.  The international trade part is represented
as follows.  Complementary imports are used by intermediary sectors
according to fixed coefficients and occur in a row.

$$m^n = \hat{b}q$$

where $m^n$ = complementary imports

$\hat{b}$   = diagonal matrix of input coefficients

$q$   = total output

Competitive imports for intermediary sector are mentioned in
brackets and are subdivided into $m^{i1}$ (quantitative constraints) and
$m^{i2}$ (qualitative constraints).  Similarly, competitive imports for
final demand ($m^f$) are subdivided into $m^{f1}$ and $m^{f2}$.  Given the state of
technical development

$$m^{i2} = M^{i2}q$$

where the coefficient matrix must be considered carefully because
technical developments in the region will reduce $M^{i2}$.  A similar analysis may be done for $m^{f2}$.

$$m^{f2} = M^{f2} \cdot f$$

Technical developments will reduce $M^{f2}$.  Income increase of
the rich may increase f and thus push $m^{f2}$ upwards, but balance of payments deficits may constrain $m^{f2}$ and may thus reduce f.  Finally, $m^{i1}$
and $m^{f1}$ are residuals, because q is capacity determined.

$$m^{i1} + m^{f1} = q^i + f - m^{i2} - m^{f2}$$

At this stage prices come to the fore possibly reducing f
rather than increasing $m^i$.

Total imports thus determined may create balance of payments
problems because exports i´e do not match with total imports i´m
(i = unity vector)

| | 1 | 2 | 3 | $q^i$ | $m^{i1}$ | $m^{i2}$ | $f^d$ | e | f | $m^f$ | $m^{f1}$ | $m^{f2}$ | q |
|---|---|---|---|---|---|---|---|---|---|---|---|---|---|
| Sector 1 | 40 (10) | 10 ( 5) | 5 (0) | 55 (15) | 8 | 7 | 100 | 70 | 170 | 10 | 8 | 2 | 200 |
| Sector 2 | 20 ( 4) | 20 (10) | 10 (0) | 50 (14) | 5 | 9 | 65 | 20 | 85 | 21 | 5 | 16 | 100 |
| Sector 3 | 10 ( 0) | 10 ( 0) | 5 (0) | 25 ( 0) | 0 | 0 | 25 | 0 | 25 | 0 | 0 | 0 | 50 |
| Sub-total | 70 (14) | 40 (15) | 20 (0) | 130 (29) | 13 | 16 | 190 | 90 | 280 | 31 | 13 | 18 | 350 |
| Value added | 120 | 40 | 30 | 190 | – | – | – | – | – | – | – | – | – |
| $m^n$ | 10 | 20 | 0 | 30 | – | – | – | – | – | – | – | – | – |
| q | 200 | 100 | 50 | 350 | – | – | – | – | – | – | – | – | – |

Table 2.5.  Input-output table for a hypothetical situation.
(competitive imports for intermediary sectors are presented in brackets)

where $m^n$ = non-competitive imports

$q^i$ = total intermediary deliveries per sector, irrespective of domestic or foreign origin

q = total output per sector

$m^{i1}$ = competitive imports for intermediary sectors, caused by quantitative constraints

$m^{i2}$ = ditto, qualitative constraints

$f^d$ = domestic final demand

e = exports

f = total final demand

$m^f$ = imports for final demand

$m^{f1}$ = ditto, caused by quantitative constraints

$m^{f2}$ = ditto, qualitative constraints

$$i\acute{}m = i\acute{}m^{n} + i\acute{}m^{i1} + i\acute{}m^{i2} + i\acute{}^{f1} + i\acute{}m^{f2}$$

Besides, at the sectoral level, import needs may be too high to be fully satisfied. It needs to be analyzed at the country level which group or sector is going to carry the burden of imbalances in production, exports and imports. It may result in more foreign aid, more foreign investment, adjustments in government income and expenditure, changes in consumption and investment, changes in prices, etc. We will touch in more detail upon such aspects of closure of the model at a later stage.

Finally, the consistency must be dealt with carefully. World exports by sector and on aggregate must equal world imports by sector and on aggregate. Systematic discrepancies arising from the system of reporting must be taken into account.

The input-output structure - At this stage we have discussed all elements of the input-output structure. They are put together in Table 2.6.

Data must be obtained from various sources because regular input-output tables do not provide all necessary information, in particular about:
- intermediate deliveries to and from the service sectors
- three types of imports - non-competitive
                          - competitive - insufficient
                                        - inadequate
- labour, by sector, by social group
- owners, by sector, by social group
- consumption, by sector, by social group
- government expenditure, by sector
- investment by origin, by destination, by social group.

Data on the above variables must be obtained from such sources as the United Nations´ Production Statistics, FAO agricultural statistics, ILO employment statistics, social accounting matrices, census data and micro studies.

For the non-value added part of deliveries to intermediate sectors, fixed input-output coefficients may be derived. Regional input-output tables may be calculated in the same way as in the ESCAP-FUGI project (cf. Chapter 3). While value added may on aggregate be assigned a fixed input-output coefficient, the composing parts of value added by sector may change over time. This has been pointed out in the discussion on production. Given sectoral labour demand, wages, net indirect taxes and depreciation, operating surplus may act as a residual rather than following total output by way of a fixed input

Table 2.6. The Input-output structure

| | Intermed sec. Total 1,2,3...20 $q^i$ | Imports intermed. sectors $m^{i1}$ $m^{i2}$ $q^i$ | Consumption labour 1,2,3,4,5,6,7 | Consumption owners 8,9,10,11 | Government con-sump-tion | Government con-sump-tion | Invest-ment 8 9 10 11 | Domes. final demand | Exports | Total final demand | Imports for final demand $m^{f1}$ $m$ $^{f2}$ $m^f$ | Total out-put $q$ |
|---|---|---|---|---|---|---|---|---|---|---|---|---|
| Agriculture | 1 2 3 | | | | | | | | | | | |
| Extractive industries | 4 5 | | | | | | | | | | | |
| Manufacturing industries | 6 7 8 9 10 11 | | | | | | | | | | | |
| Construction | 12 13 | | | | | | | | | | | |
| Electricity, gas | 14 | | | | | | | | | | | |
| Water | 15 | | | | | | | | | | | |
| Merchandise, trade, commerce | 16 17 | | | | | | | | | | | |
| Formal services | 18 19 | | | | | | | | | | | |
| Informal services | 20 | | | | | | | | | | | |
| Total intermediate | | | | | | | | | | | | |
| Non-competitive imports | | | | | | | | | | | | |
| Total non-value added | | | | | | | | | | | | |

Table 2.6. continued.

| | | |
|---|---|---|
| Labour, rural | 1a-1 | |
| | 1b-2 | |
| | 2b-3 | |
| urban | 1a-4 | |
| | 1b-5 | |
| | 2bc-6 | |
| | 3bc-7 | |
| Total labour | | |
| Owners, rural | 2a-8 | |
| | 3 -9 | |
| urban | 2a-10 | |
| | 3a-11 | |
| Total owners | | |
| Depreciation | | |
| Net and indirect taxes | | |
| Total value added | | |
| Total output | | |

coefficient. Apart from the value added sectors, input-output coefficients will not change dramatically owing to technical progress and resource constraints. The latter phenomenon may result in limited growth for a certain sector if imports of required resources are not sufficiently feasible because of financial reasons or limited availability.

Determination and influence of prices and wages in general will be discussed at a later stage. Here we will pay attention to the effect of prices on input-output coefficients. Since input-output tables are in value terms, changes in relative prices will require adjustment in input-output coefficients. However, it would be too cumbersome to adjust the current year´s coefficients simultaneously with the current year´s price changes. To reduce the complexity and the computational load of the model, input-output coefficients will be considered predetermined for the current year and adjusted for the following year. Current world developments heavily necessitate emphasis on this aspect for the energy sectors (sectors 4 and 5) and their influence on balance of payments because energy is one of the imported commodities for most countries in the ESCAP region.

Wages, prices and exchange rates - Although a complete general equilibrium model is hard to apply in the model system proposed here, wages, prices and exchange rates should take care of at least part of the clearing of the market. This particularly concerns supply and demand of labour, supply and demand of products (goods and services) and supply and demand of capital.

Wages are different for rural groups and for urban groups, partly because of the higher cost of living in urban areas. Wages are interdependent with prices, specifically food prices. Low wages in the rural areas are the basis for low food prices and for international competition in export of primary commodities. Not allowing rural wages and food prices to increase substantially makes it possible to keep wages in the urban areas low and thus improve profitability and international competitiveness.

It is suggested above to accept the fixed share of *aggregated* labour, profit, and other value added components, giving room for the various value added components to adjust simultaneously within this boundary. The adjustment mechanism is triggered off by changes in labour supply and demand, resulting in equilibrium wage rates which are effectuated in the informal tertiary and quartary sectors together with the agricultural sector.

wage rates in the informal tertiary sector should reflect the pressure on this sector exerted by residual workers pushed out of

agriculture and pulled into the urban sector. Their total aggregated
income should equal the purchasing power reflecting demand for tertiary
and quartary informal services; these services being produced without
labour-saving capital equipment. The urban pull should be dominated
by the absorption capacity and wage rates of the large-scale urban
sectors.

Prices and wages are basically interrelated through the
input-output structure. For some (primary) sectors the domestic price
level is based on world market prices for the commodities concerned.
Duties and subsidies by the government may act as factors of adjustment. Similarly, import prices are exogenous to the countries concerned
and may be affected by the government as well. In each case it needs
to be decided to what extent domestic production prices, final demand
prices and intermediate delivery prices are related to that sector´s
import prices. Homogeneity of the sector and the functioning of competitive imports will play an important role. In most cases it must be
assumed that final demand prices and prices of intermediate deliveries
are the same and may even equal production prices.

Modelling the exchange rate is very complicated. It is very
hard to draw up a model describing a relationship between the exchange
rate and other variables such as balance of payments deficits or surpluses, inflation and policy elements such as the need to attract investment and savings from abroad. On the other hand, the exchange rate
will have its influence on prices and thus on the international competitive position and on domestic prices.

Income distribution - Concerning income distribution, three
aspects are important:
- how is income distribution measured;
- which variables influence income distribution;
- what is the role of income redistribution towards other variables?
We shall discuss these aspects in the context of our model.

Income distribution in many studies is measured as income
distribution by size through such statistics as the Gini-coefficient
or Shannon´s entropy measure. The basis for these statistics is a
division of households into income groups. Using average income per
group and assuming (lognormal) distribution function, income distribution within groups and for total population can be derived.

Carrying out the above analysis in the context of our model
implies doing a completely separate job. Endogenizing such work in
the model by linking it to income by social groups (cf. Table 2.4.)
and partly disaggregated by groups of sectors means a huge task, the
benefits of which are not completely clear compared to the costs

involved. Studies have shown that the size distribution of income is exceedingly stable, represented by a change in the Gini coefficient which is rarely more than a few percentage points cf. Adelman et al., (1979). Besides, it is concluded that the relative position of various socio-economic groups is more sensitive to policy changes than is the overall distribution.

Average income by social groups and income distribution between social groups may be derived from the model and its input-output structure by looking at value added, the bottom left part of Table 2.6. Special attention should be paid to average income for such groups as the rural landless unskilled labourers and the urban unskilled in the informal service sector. The level of average income for these groups may be as good a statistic to concentrate on as any other statistic for policy reasons.

Looking at income distribution in this way, it is important to find the most relevant variables influencing income for the above mentioned lowest income groups. Basic variables are wages/prices, production (functions), government policies (subsidies, taxes, basic needs provisions), population growth (labour supply, etc.) and labour demand by other sectors. Thus, income distribution is determined by some variables in particular and by the model as a whole in general.

Which are the effects of income redistribution?

The most important effects are in the area of consumption (and savings). Consumption will move to a composition with a lower import content and a higher labour content, thus in turn influencing income distribution positively.

Influences of developments in non-ESCAP countries - Classification of the world by regions in Table 2.2., shows seven regions (regions 14-20) which are relevant for the subject matter of this model only from the international trade and capital flows side. Perhaps, with the exception of Japan, it may suffice to postulate exogenous (scenarios of) growth rates and handle these regions rather similarly to the ESCAP-FUGI model. Treatment of Asian Centrally Planned Economies will create problems and will require solutions, which must be focussed upon separately at a later stage.

Bibliography

Adelman, I., M.J.D. Hopkins, G.B. Rodgers, R. Wéry, S. Robinson, (1979):
    "A Comparison of Two Models for Income Distribution Planning",
    Journal of Policy Modeling, Vol. I, no. 1.

Asimakopoulos, A. (1977): "Profits and Investment, a Kaleckian Approach",
    in: Harcourt, G. (ed.), The Micro-economic Foundations of
    Macro-economics, MacMillan, London.

Berry, R.A. and Cline, W. (1979), Agrarian Structure and Productivity
    in Developing Countries, Baltimore and London.

Blitzer, R., P.B. Clark, L. Taylor (1975): "Economy-wide Models and
    Development Planning", New York, Oxford University Press.

Fanjul, O. (1979): "A Dynamic Multisectoral Model of Income Redistribu-
    tion, Employment and Growth", 7th United Nations´ Input-Output
    Conference, Innsbruck.

Ghosh, A., A.K. Sengupta (1979): "Income Distribution and Production
    Structure in Input-Output Framework," 7th United Nations´
    Input-output Conference, Innsbruck.

Hopkins, M.S.D., R. Van Der Hoeven (1979): "Economic and Social Factors
    in Development: A Socio Economic Framework for Basic Needs
    Planning", World Employment Programme Research Working Paper,
    International Labour Office, Geneva.

Kaya, Y., A. Onishi, et. al. (1977): "Report on Project FUGI, Future
    of Global Interdependence", 7th IIASA Global Modeling Con-
    ference, Laxenburg, Austria.

Kaya, Y., A. Onishi, S. Abe, H.P. Smit (1979): "Long-term Projections
    of Economic Growth in ESCAP Member Countries", 7th United
    Nations´ Input - output Conference, Innsbruck.

Lewis, W.A. (1954): "Unlimited Supplies of Labour", in: Manchester
    School.

Lipton, M. (1977): Why Poor People Stay Poor, Temple Smith, London.

Manne, A.S., T.E. Weisskopf (1970): "A Dynamic Multisectoral Model for
    India, 1967-1975", in: A.P. Carter and A. Brody, Applications
    of Input-output Analysis, North Holland Publishing Company.

Melo, J. De, S. Robinson (1979): "Income Distribution and Trade Policy
    in a Multisector Planning Model", 7th United Nations´ Input-
    Output Conference, Innsbruck.

Onishi, A., Y. Kaya (1980): "Long-term Projections of the Economies of
    ESCAP Countries"; paper prepared for ESCAP, Bangkok.

Panchamukhi, V.R. (1974): "A Multisectoral and Multicountry Model for
    Planning ECAFE Production and Trade", in: K.R. Polenske and
    J. Skolka, Advances in Input-Output Analysis, North Holland
    Publishing Company.

Paukert, F., J. Skolka and J. Maton (1979): "Income Distribution by
    Size, Structure of the Economy and Employment: A Comparative
    Study for Four Asian Countries", 7th United Nations´ Input-
    Output Conference, Innsbruck.

Pyatt, G., J.I. Round (1977): "Income Distribution and Input-Output:
    An Analysis for Sri Lanka", Warwick Economic Research Papers,
    University of Warwick, Coventry.

Rodgers, G.B., M.S.D. Hopkins, R. Wéry (1978): "Population, Employment
    and Inequality, BACHUE Philippines", International Labour
    Office, Saxon House.

Schultz, T.W. (1955): "The Role of Government in Promoting Economic Growth", in: L.D. White (ed.), The State of Social Sciences, London.

Vos, R.P. (1979): "The State and Economic Development", Department of Economics, Free University, Amsterdam.

# MODELLING POLICY CONSEQUENCES AND EVALUATION PROCESSES
## USING THE "DEDUC" NONNUMERICAL PROGRAM SYSTEM

Hartmut Bossel*


Gesamthochschule, D 35 Kassel, F.R. Germany and
Institut für Angewandte Systemforschung und Prognose (ISP),
Medizinische Hochschule, Haus W3G, D 3 Hannover, F.R. Germany

## Abstract

The reality which global models are supposed to model contains many crucial processes (ecological and social development, human
decision-making, etc.) which cannot be properly modelled by numerical
methods and must be represented by the logical processing of qualitative, nonnumerical information. We have developed, and are using for
these purposes, the DEDUC nonnumerical processing system based on the
predicate calculus. Elements of the system are briefly described. It
has proven itself very useful and reliable in processing large amounts
of qualitative data for a) deducing impacts and consequences of events,
actions or policies (environmental impact assessment is a typical
application) and b) for obtaining impact evaluations from the point of
view of given social actors in order to obtain reliable assessments of
likely attitudes and behavioral trends concerning given events, actions
or policies. Supplementing quantitative global models with qualitative
models of relevant non-quantifiable processes would improve their
analytical and predictive quality significantly.

## Global Modelling and Qualitative Systems Analysis

Quantitative modelling can only cover certain aspects of
reality, i.e., those for which quantitative data can be obtained by
measurement or inference, and for which natural laws are applicable

## Acknowledgements

The author gratefully acknowledges the generous support of the Volkswagen Foundation, Hannover. The present paper uses some material
("Elements of the DEDUC system") previously published in the Journal
of Policy Analysis and Information Systems [1].

* Paper based partially on unpublished research by K.F. Müller-
  Reissmann, B. Hornung, and R. Schiefer of ISP.

(including, for example, parts of the biological and social sciences) and can be cast into quantitative form. However, important aspects of the systems represented in global models are of a qualitative nature. One aspect is the fact that in these systems human actors usually play a more or less decisive role. While, of course, subject to natural laws, humans observe the developments about them, combine their observations with their personal knowledge and beliefs, draw conclusions, subject these to an evaluation process using their own normative criteria, and act on the basis of the evaluation results. The outcome of this nonnumerical, qualitative reasoning process (which may not even be fully rational, i.e., it could be dominated by affects and emotions) may - and often does - significantly affect the development of the global system[2, 3, 4, 5, 6].

Another aspect is the fact that with respect to some parts of a global system certain causal relationships may be known very well indeed although the variables themselves and the relationships between them defy quantification. These "soft" or "fuzzy" variables and relationships may even be crucial determinants of the system´s behavior, and their deletion would invalidate the whole modelling effort. Ecological systems provide many examples of such variables and relationships[7, 8, 9].

Example: In an ecological system of "great diversity" and "robustness", "strong environmental stress" might lead to the temporary reduction in the number of individuals of some species, but the system would recover its original "stability" "after some time". In a system having "little diversity" or "poor robustness" the same "strong environmental stress" might cause extinction of some key species and the establishment of an ecological system quite different from the original. The expert can use this (or other qualitative knowledge) to provide an accurate prediction of system behavior even though these "soft" variables and relationships may be impossible to quantify and model numerically.

Many attempts have been made to cast such relationships into a quantitative form permitting numerical computation. They are a regular feature of most system dynamics models ("policies"). While often these may be reasonably valid representations of policy formation processes or qualitative relationships, it must be realized that the method of representation is basically inappropriate. "Fuzzy systems analysis"[10] can successfully deal with the qualitative aspects of such relationships, but its application to the often very intricate relationships of deductive reasoning or ecological or social impact analysis becomes awkward to the point of impracticality.

The common features of those aspects of reality requiring a valid qualitative description in global models are
- predicates representing qualitative states of objects (e.g. "polluted");
- objects, often representing merely qualitatively defined entities (e.g., "environment");
- statements (premises) describing states of given objects (e.g. "polluted (environment)");
- implications representing the relationships between predicates for different objects (i.e., between the states of different objects) (e.g. "*if* polluted (environment) *and* available (money) *then* installation (pollution control equipment)");
- object structures for establishing relationships among objects (e.g. "air, rivers, sea, ecosystems *is* environment").

With these elements it is possible to construct qualitative simulation models from which qualitative conclusions can be deduced. The following table lists the correspondences between quantitative (dynamic) simulation models and qualitative models.

| general description | quantitative description | qualitative description |
|---|---|---|
| state variable | variable | object defined by object intersection* |
| state | value | predicate |
| structure | functional relationships | implications |
| initial condition | initial values | premises |
| boundary conditions | boundary values | premises |
| (scenario) parameters | parameters | premises |
| resulting state | output vector | conclusions |

* e.g. (copper, electric motor, Poland, today) = the copper which is required for the construction of electric motors in Poland today.

The appropriate way to deal with this type of qualitative systems description is the predicate calculus. As the result of a compromise between the full powers of the predicate calculus and the practical limitations of simulation efforts for policy analysis, F. Rechenmann and K.F. Müller-Reissmann have developed the DEDUC program system whose basic features are briefly described in the following[1, 11, 12].

## Elements of the DEDUC System

DEDUC is an interactive computer program system for the deductive processing of nonnumerical information ("concepts") supplied by the program user. The system contains only the procedures for storing, processing and outputting information. All procedures are written in FORTRAN IV and use nonnumerical list-processing (only the certainty factor and loading factor calculations are numerical). The system will be explained by applying it to a simple example of impact analysis.

The elements of the DEDUC-language are <u>objects</u> characterized by arbitrary names (normally abbreviated to 8 alphabetic characters) such as

RESOURCE, PRODUCT, FRANCE, NEAR-FUTURE

and <u>predicate statements</u> having the general form

"predicate" ("object 1", ... , "object n")

(up to a maximum of n = 8, i.e., 8-place predicates). Single predicates may be written as either

"predicate" ("object") or "object" "predicate".

Examples:

REQUIRED (RAW MATERIAL, PRODUCT, COUNTRY, TIME)

"A (certain) raw material is required for the production of a product in a country at a given time"

EXPENSIVE (HOUSE) or equivalently:

HOUSE EXPENSIVE

"(The) house is expensive"

Note that it is up to the user to define an unambiguous meaning for each predicate statement. The statement by itself may sometimes not be self-explanatory.

Concepts and their relationships may be entered in one of three forms: (1) object structure definitions, (2) implications, and (3) premises.

<u>Object structure definitions</u> serve for entering classificatory "world knowledge" in the form of objects or object classes and their respective subclasses and specimens. The object structure definitions command is composed of the object structure list on the left-hand side, the <u>IS</u> keyword and the global object structure name on the right-hand side:

•    "object structure list" <u>IS</u> "global object structure name";
rather complex object structures can be constructed by proper use of parentheses.

Examples:

RAW-MATERIAL (COAL, OIL, IRON), FERTILIZER, WHEAT IS RESOURCE;

"Coal, oil, iron are raw materials; raw materials, fertilizer, wheat are resources"

IND.PRODUCT (FERTILIZER), AGR.PRODUCT (WHEAT), BREAD IS PRODUCT;

"Fertilizer is an industrial product; wheat is an agricultural product; industrial products, agricultural products and bread are products"

IND.COUNTRY (GERMANY, USA), DEV.COUNTRY IS COUNTRY;

"Germany and USA are industrial countries; industrial countries and developing countries are countries"

PRESENT, NEAR-FUTURE, FAR-FUTURE IS TIME;

"The present, the near future, and the far future are (points in) time"

Implications provide a means for entering relationships between predicate statements. These IF - THEN relations are used to describe causal, consecutive, conditional, or identifying world knowledge in the form of concepts composed of logical relationships between predicate statements. Implications have the general form

IF "predicate expression" THEN "predicate list" "CF"

The predicate expression on the left-hand side is composed of predicate statements connected by the logical operators AND, OR and NO and parentheses. Parentheses in the predicate expression may be deleted, provided the precedence ordering NO, AND, OR is observed. The predicate list on the right-hand side consists of a list of negated or non-negated predicate statements separated by commas. The certainty factor CF may be assigned any value between 0 and 100 (per cent). It is a (subjective) measure of the certainty that the statement is true. The program automatically assigns a certainty factor of 100 if no number is provided. It also automatically generates all logically permissible inverse implications for each implication entered, unless told not to do so.

Example:

IF    (SCARCE (RESOURCE, COUNTRY, TIME)

OR    EXPENSIVE (RESOURCE, COUNTRY, TIME))

AND   REQUIRED (RESOURCE, PRODUCT, COUNTRY, TIME)

THEN  EXPENSIVE (PRODUCT, COUNTRY, TIME);

"The following holds for all countries at all times: if a resource becomes scarce or if its price increases, and if this resource is required for the production of a certain product, then the price for this product will increase".

Premise definitions serve to enter premises, i.e., single "true" predicate statements from which the deduction routine then attempts to generate new "true" statements via the implications. Their general form is

PREM "predicate list"

Example:

PREM    SCARCE (OIL, COUNTRY, NEAR-FUTURE),
            REQUIRED (OIL, FERTILIZER, COUNTRY, NEAR-FUTURE),
            REQUIRED (FERTILIZER, AGR.PRODUCT, IND.COUNTRY, TIME),
            REQUIRED (WHEAT, BREAD, GERMANY, TIME);

"Oil will become scarce in all countries in the near future. Oil will be required for the production of fertilizers in all countries in the near future. Fertilizers are required for the production of agricultural products in industrial countries. Wheat is required for the production of bread in Germany".

The object specification and execution command

CONCL "object list";

starts the deduction process with respect to all implications and premises previously entered. At the same time it causes the program to select from the set of generated conclusions those conclusions whose objects allow a reduction to the objects given in the object specification. Only these conclusions are presented as output.

Impact analysis - We use the examples of object structures, implications, and premises listed in the examples above to demonstrate a (simplified) impact analysis using the DEDUC system. If the program user is interested in generating consequences concerning the "near future" he starts the deduction process by entering

CONCL NEAR-FUTURE

Using the available concepts, DEDUC generates the following output sequence:

EXPENSIVE (FERTILIZER, COUNTRY, NEAR-FUTURE)
EXPENSIVE (AG.PRODUCT, IND.COUNTRY, NEAR-FUTURE)
EXPENSIVE (BREAD, GERMANY, NEAR-FUTURE)

Note that unless finely scaled predicates are used, the impact analysis will only indicate qualitative effects, not quantitative magnitudes of the effects.

Some additional features of DEDUC - In a complex deduction process it becomes almost impossible for the program user to realize how a certain conclusion may have been generated. The HOW command allows him to trace the deduction process all the way from the final conclusion through the different implications and premises used.

The subjective certainty with which an actor believes a

concept to be true has little to do with the objective probability of it actually being true. Nevertheless, it is this subjective certainty which enters the reasoning process and must be properly considered. This processing must account for certain fundamental properties of uncertainty which distinguish it from probability; hence the application of the rules of probability would not be appropriate and we have therefore implemented a distinctly different certainty calculus[1]. Conclusions having a low certainty factor can be deleted from further consideration by specifying an appropriate CF cutoff value.

In evaluation processes (see below), an impact may load on different orientors to a different degree. These differences are accounted for by assigning the appropriate loading factors LF to orientor impact implications. The program takes account of these loading factors and orientor weights in computing overall orientor impacts.

The program can also deal with a time sequence (TIME, TIME+, TIME++, ... ). This allows some amount of dynamic processing, including the representation of feedback effects.

The program prints a number of self-explanatory messages and questions to aid the user, such as warnings pointing to inconsistencies in the conceptual system and reminders to supply additional information which may be relevant to the problem, but which the user forgot or did not recognize as being important.

## Use of the DEDUC System for Impact and Consequence Analysis

A first important task to which the DEDUC system can be applied is the field of impact and consequence analysis. The task may be described as follows:

Given: diverse knowledge about all relevant aspects of a given system;
an initial system state;
a contemplated or expected action, event, or policy input, i.e., a control or disturbance input to the system.

Find: impacts and consequences of the action, event or policy.

The DEDUC example in the previous section illustrates the approach. A more realistic impact analysis for policy applications would normally contain hundreds of objects and several hundred implications representing all relevant knowledge about the subject area. Consequently, a very large number of impacts will be deduced by the system. (The user will normally minimize the amount of output by specifying only those objects for which he wants to see the impacts).

It is obvious that the deduction of impacts and consequences can only be as good as the knowledge represented in the object and implication structure of the DEDUC implementation.  Also, it should be stressed that the conclusions are qualitative and will often only give crude hints concerning the actual magnitude of impacts that will occur.  If exact quantitative data for expected impacts are important, quantitative simulation will have to supplement the analysis.  The qualitative impact analysis however will ensure that - to the best of current knowledge as incorporated in the knowledge module - all possible consequences of an action are deduced and their possible impacts listed. This feature is of considerable value in environmental, social and political impact analysis, where many important variables are of a qualitative nature and where the construction of reasonably reliable quantitative models is out of the question.

The advantages of qualitative modelling with DEDUC are the following:

- All information considered relevant can be included (quantitative information must be represented "verbally" in the form of "small increase", "large decrease", etc.).

- The number of objects and implications can be very large without causing any particular problems.

- Due to the number of objects and their possible intersections as arguments of predicates, the number of "state variables" can be extremely large.

- Despite the huge number of conclusions drawn internally, the user can restrict the output to only those impact areas in which he is interested.  He can also retrace the chain of individual conclusions (using the HOW-command) in order to understand how and why they were generated.

- The program system permits a certain amount of dynamic (and feedback) analysis by using TIME as an object in the predicate argument.

- The DEDUC formalization approximates actual human reasoning and primitive language expression and is therefore quickly learned by users.

## Use of the DEDUC System for Impact and Policy Evaluation

While the qualitative modelling of impacts is an important though straightforward application of DEDUC, it was originally constructed with another application in mind: the simulation of the cognitive and normative processes of human reasoning and decision-making. In these processes, the impact analysis as described above is followed by a process of impact evaluation in which the impacts are mapped onto relevant evaluation criteria (orientors) in order to determine whether, and how strongly, they are affected in a positive or negative sense. The decision for or against a certain policy can then be expected to be determined by a comparison of the sum of the weighted impacts on the individual orientor criteria.  The task is described as follows:

Given: a certain policy, action or event;

the "world knowledge" of an actor;

his normative criteria and their respective weights (orientors);

an initial system state.

Find: the contributions of the likely impacts to orientor satisfaction or dissatisfaction, and hence the actor´s evaluation of the ensuing situation and his overall (positive or negative) reaction.

Impact evaluation - For impact evaluation, the results of the impact analysis in the knowledge module are transferred to the orientation module[13, 14].  The orientation module consists of a more or less developed hierarchy of orientors (goals, values, basic orientors, etc.) which are entered as object structures, premises and implications.  The orientation module for our simple example could have the following components:

PHYS.EXISTENCE, SECURITY, FREEDOM, EFFICIENCY,

ADAPTIVITY IS BASIC-ORIENTOR;

"Physical existence needs, security, freedom of action, efficiency and adaptivity are basic orientors".

HEALTH, NUTRITION, SOCIAL-SECURITY, ... IS VALUE

"(Adequate) health, nutrition, social security, etc., are values".

Implications and object structures must be used to represent the relationships between state descriptions (the output of the impact analysis) and the orientors:

BREAD, POTATO IS BASIC-FOOD;

IF   EXPENSIVE (BASIC-FOOD, COUNTRY, TIME)

THEN THREAT (NUTRITION, COUNTRY, TIME);

"If (in a given country at a given time) a basic food is expensive, then the (value) nutrition is threatened".

Since the impact evaluation is to be made by reference to the basic orientors, appropriate general statements relating the threat to an important value to corresponding effects on basic orientors must be made:

<u>IF</u>   THREAT (VALUE, COUNTRY, TIME)

<u>AND</u>  IMPORTANT (VALUE, BASIC-ORIENTOR, COUNTRY, TIME)

<u>THEN</u> THREAT (BASIC-ORIENTOR, COUNTRY, TIME);

"If (in any country at any time) a value is threatened, and if this value is important to (the satisfaction of) a basic orientor, then that basic orientor is threatened".

<u>PREM</u> IMPORTANT (NUTRITION, PHYS.EXISTENCE, COUNTRY, TIME);

"Satisfaction of the value nutrition is important for (the satisfaction of the basic orientor) physical existence".

With these concepts in the orientation module the impact evaluation can now be made after transferring the results of the impact analysis and continuing the deduction process. The result of the impact analysis

EXPENSIVE (BREAD, GERMANY, NEAR-FUTURE)

gives rise to the following evaluation results:

THREAT (NUTRITION, GERMANY, NEAR-FUTURE),

THREAT (PHYS.EXISTENCE, GERMANY, NEAR-FUTURE)

This example merely illustrates the approach. Realistic applications require a much more detailed orientor structure as well as a more differentiated description of the relationships (such as "mild threat", "threat", "strong threat", "somewhat important", "important", "very important"; or even finer graduations). This can be accomplished by the use of predicate variables[1].

The output of the orientation module is a vector giving the degree and certainty of the different orientor impacts on the different systems considered and at different points in time. This impact evaluation therefore provides a very differentiated picture of the consequences of a given state or policy in orientation space. This orientation guides the search for a satisfactory policy. For a numerical approach using the same concepts, see an earlier publication[15].

Practice has shown that a "verbal" evaluation as described is clumsy and probably not accurate enough for a realistic description in many cases. We have therefore incorporated into DEDUC a numerical loading factor calculus and given the user the possibility to assign his own orientor weights in order to better represent his personal preferences (or those of the actor he is simulating). The program then computes the positive and negative weighted contributions to each orientor separately and presents the results on a bar chart. By comparing respective bar charts, differences in evaluations between actors and between alternatives become obvious. Where the reasons for these differences are not obvious, they can be determined by tracing back through the evaluation process using the HOW-command.

## Applications

### Environmental impact analysis: the incorporation of physical characteristics, legal norms, economic considerations, and evaluation

The object of one present study (financed by the German Federal Environmental Agency) is the development of an instrument which will provide a comprehensive impact analysis of existing and proposed legislation in the energy field. The major components of the instrument are:

- a program system which permits the quantitative modelling of an arbitrary energy supply system and the computation of the resulting energy flows and environmental consequences;

- comprehensive physical data sheets for each technical process (e.g. mining, transportation, conversion, use of energy);

- comprehensive legal data sheets for each technical process and for each of its physical consequences (e.g. building codes, environmental regulations, tariffs, etc.);

- a DEDUC knowledge system incorporating all relevant knowledge about the physical characteristics, legal ramifications and economic consequences of the different technical components of the energy supply system and containing the impacts on orientors in order to allow impact evaluation.

While different applications are possible, the system would normally be used as follows:

1 - The consequences of likely changes in the energy supply structure due to a new legal norm are first determined using the DEDUC knowledge module and possible additional expert knowledge not incorporated in this module.

2 - This (qualitative) information is translated by the user into quantitative inputs to the (numerical) model of the energy supply system.

3 - The simulation model of the energy supply system computes the resulting energy flows and environmental consequences.

4 - The information on the deployment of technologies is used to deduce the environmental and legal impacts of the different energy technologies from the DEDUC knowledge module.

5 - These impacts are evaluated in the DEDUC evaluation module.

6 - The user compares the evaluation result with his intentions and, if necessary, changes the legal norms in order to achieve better results (start again from "1").

It is obvious that this procedure is far from being an

automatic legal norms generator. However, it can guide and aid the norms formulation process, assure a reasonably complete impact assessment process ("to the best of our knowledge"), and provide a comprehensive evaluation of proposed changes.

## Cognitive systems analysis: predicting responses of social actors

Unless we are talking about a single person, the "social actor" standing for a political party, a labor union or any other social group or organization can only be a theoretical fiction. However, for the analysis of the development of social systems, the representation of the most influential groups or organizations in the form of prototypical actors is often justified and may provide valid conclusions at least about the basic development trends which must be expected. Perhaps the most important determinants of the behavior of social actors are (a) their knowledge and beliefs, and (b) their orientors (evaluation criteria) and their respective weights. Both of these can be represented by a DEDUC formalization, which can then be used to determine the response of the social actor to policies, actions or events (entered as premises). The difficult part is the development of an accurate cognitive and normative representation of the social actor. Since "the" actor does not exist, and the representative opinion leaders may often act on the basis of cognitions or orientors of which they are not themselves aware (or which they do not want to admit), the construction of the proper knowledge and orientation modules must rely partly on content analysis, partly on expert knowledge, partly on the intuition of the researcher. Despite the objections which can be raised against such a representation of the social actors, we feel that the method can be used to obtain reasonably reliable assessments of the behavioral trends of social actors. However, its limitations must be kept in mind.

We have so far constructed knowledge and orientation modules with respect to environmental, energy and economic issues for the West German Social Democratic Party (SPD), the ecological movement, and the Christian Democratic Party (CDU). These have been used to deduce responses to certain policy issues, e.g., the increased construction of nuclear plants, introduction of organic agriculture, or decentralization of the energy system.

Descriptions of the policy issues, events or actions to be tested are entered as premises into the knowledge module of the social actor whose response is to be analyzed. The knowledge modules of the three actors mentioned contain each several hundred objects and several hundred implications, each with a corresponding certainty factor

measuring the strength of belief in a given implication.  The impacts
which follow on the basis of the available knowledge are deduced and
then transferred into the orientation module.  The orientation module
contains the basic orientors[13] (existence, security, freedom of action,
efficiency of control, adaptivity) and some 100 derived orientors on
four hierarchical levels in all.  The choice of these orientors is the
result of an extensive theoretical and empirical analysis[16].

In the orientation module, the impacts are first mapped onto
the lowest level orientors (applying the appropriate loadings as given
in the implications), and the corresponding weighted orientor impacts
are then aggregated to higher level orientor impacts.  The evaluation
results on the level of the socially relevant orientors (second hierarchical level from the top) are then presented in graphical form.
These orientors are: health, material welfare, self-actualization,
personal freedom, freedom of action of subsystems, freedom of action
of the total system, ecological efficiency, economic efficiency, control efficiency, system safety, supply security, flexibility, capability
for innovation, social responsibility, future responsibility.

Figure 1 presents the bar charts for two different actors
(SPD and ecologists) evaluating an identical policy ("increased construction of nuclear plants in F.R. Germany").  It is obvious that one
actor ("ecologies") strongly rejects this policy, while the other, SPD,
arrives at a mixed evaluation from which his actual behavior would be
more difficult to predict.  These results accurately describe the
current real situation.

The conclusions generated by the DEDUC process concerning
many other issues have always been plausible and are in general agreement with intuitive expert assessment or observations.  We therefore
feel that carefully constructed knowledge modules and orientation modules of prototypical social actors, if cautiously used, can provide
reasonably accurate assessments of responses to given policies, events
or actions.

By analyzing the chains of deduction using the HOW-command,
one can trace the causes for the different results of the impact
analysis and the impact evaluation.  These are:

- a different conceptual structure (objects, implications,
  and premises) in the knowledge modules of the different
  actors;
- a different conceptual structure in the respective orientation modules;
- different certainty or loading factors attached to identical concepts;

114

VALUE | WT. | CF- | CF+ | WEIGHTED, AGGREGATED EVALUATION

```
                              -300    -200    -100      0     100    200    300
                               |       |       |        |      |      |      |
GESUNDH  | 60 | 55 |  0 |                                ---0
WOHLST   | 80 | 35 | 53 |                        ----------0++
SVWIRK   | 30 | 55 |  0 |                                --0
PERSFR   | 40 | 55 |  0 |                                ---0
SUBSYFR  | 20 | 55 |  0 |                                 -0
GESYFR   | 40 | 40 |  0 |                                ---0
OEKLEFF  | 10 | 53 |  0 |                                  0
OEKNEFF  | 60 | 28 | 85 |                             -----0++++
STEFF    | 50 | 53 |  0 |                                ----0
GFSI     | 60 | 40 | 21 |                    -------------0+
VSORGSI  | 70 | 52 | 21 |                        ---------0++
FLXIB    | 20 | 53 |  0 |                                ---0
INNOVF   | 20 | 60 |  0 |                                 --0
SOZVAW   | 50 | 38 |  0 |                                ---0
ZKFTVAW  | 30 | 52 |  0 |                                 --0
```

OK:

(a) moderate opponent to nuclear power plant construction

Figure 1. Comparison of the impact evaluations of two different social actors concerning an identical policy input (construction of nuclear power plants). (a) "Social Democratic Party", (b) Ecological Movement.

```
VALUE      WT.  CF-  CF+   WEIGHTED, AGGREGATED EVALUATION

                       -300    -200    -100      0      100     200     300
                         :       :       :       :       :       :       :

GESUNDH   | 90 | 66 |  0 |------------------------0
WOHLST    | 40 | 45 |  0 |               --------0
SVWIRK    | 70 | 77 |  0 |       ----------------0
PERSFR    | 70 | 77 |  0 |         --------------0
SUBSYFR   | 70 | 86 |  0 |             ----------0
GESYFR    | 60 | 90 |  0 |                -------0
OEKLEFF   | 90 | 51 |  0 |                 ------0
OEKNEFF   | 50 | 48 | 90 |                     --0++++
STEFF     | 50 | 82 |  0 |             ---------0
GFSI      | 70 | 81 |  0 |    ------------------0
VSORGSI   | 70 | 45 |  0 L------------------------0
FLXIB     | 40 | 57 |  0 |               -------0
INNOVF    | 40 | 36 |  0 |                 -----0
SOZVAW    | 80 | 45 |  0 |         ------------0
ZKFTVAW   | 90 | 45 |  0 |             --------0

OK:
```

(b) radical opponent to nuclear power plant construction

Figure 1 continued.

> - different weights of orientors, and different social and
> future horizons.

Putting it in more familiar terms: cognitive systems analysis
by nonnumerical simulation shows quite clearly the three major reasons
why people arrive at different conclusions concerning the same issue:
1) differences in knowledge, 2) differences in the strength of beliefs
attached to uncertain knowledge, 3) differences in their ethical framework. One conclusion from this is that conflicts on policy issues
cannot be resolved by more or better information alone, as long as
ethical positions differ[17]. Another possible difference is due to the
likely modification of purely rational reasoning by psycho-logical
mechanisms. We have not yet attempted to implement this aspect.

## Summary and Conclusions

Reality is determined to a great extent by qualitative relationships and by the results of human reasoning processes which use
qualitative information almost exclusively. Global models in particular are confronted with the problem of representing this information
and the corresponding information processing properly. Its translation
into a (small) number of quantitative relationships ("policy functions",
etc.) is inappropriate in principle, although it may be useful in some
cases.

In order to deal with large amounts of qualitative information in a proper and practical way, we have developed, and are now
routinely applying the DEDUC program system based on the predicate
calculus. The DEDUC formalization approximates (primitive) natural
language and human reasoning processes. Given the description of an
event, policy or action, the DEDUC process will generate all the relevant conclusions which can be generated from the knowledge stored.

The DEDUC system is particularly useful for 1) impact assessment, and 2) impact evaluation from the point of view of a given
(social) actor. In the first case, the likely consequences of a proposed action are determined using all available knowledge concerning
the subject. We have used this approach for comprehensive environmental impact assessment. DEDUC even allows the consideration of feedback processes and therefore permits a crude qualitative dynamic simulation. In the second case, the impact obtained is mapped onto the
orientors (evaluation criteria) of the actor being considered. The
evaluation result provides an indication of the attitude the actor is
likely to assume concerning the event causing the impact. We have
applied this latter approach to assessing the attitude of different

political actors concerning issues in energy and economic policy. The experience has been very encouraging, and we feel now that nonnumerical processing of information will provide an important contribution towards improving the performance and validity of global models. Our experience indicates that the direct coupling of quantitative and qualitative simulations should be avoided and that the results of the qualitative impact analysis and impact evaluation should rather be interpreted by the user and used as guidelines for the development of proper exogenous parameter inputs (scenario and policy parameters) for global models.

## References

1. Bossel, H., Müller-Reissmann, K.F.: Simulation of the Cognitive Processes of Policy Analysis. Policy Analysis and Information Systems, Vol. 3, No. 1, July 1979, pp. 1-25.

2. Axelrod, R.: Structure of Decision - The Cognitive Maps of Political Elites. Princeton University Press, Princeton, N.J., 1976.

3. Carrol, J.S., Payne, J.W. (eds.): Cognition and Social Behavior. Erlbaum, Hillsdale, N.J., 1976.

4. Kirsch, W.: Entscheidungsprozesse. Gabler, Wiesbaden, 1970, 1971 (3 vols.).

5. Neisser, U.: Cognition and Reality. Freeman, San Francisco, 1976.

6. Bossel, H.: A Modelling Framework for Societal Systems. In H. Bossel (ed.): Concepts and Tools of Computer-assisted Policy Analysis. Birkhäuser, Basel, 1977, Ch. 4, pp. 162-179.

7. Ehrlich, P.R., Ehrlich, A.H., Holdren, J.P.: Ecoscience - Population, Resources, Environment. Freeman, San Franscisco, 1977.

8. Kormondy, E.J.: Concepts of Ecology. Prentice-Hall, Englewood Cliffs, N.J., 1976.

9. Holling, C.S.: Adaptive Environmental Assessment and Management. Wiley-Interscience, New York, 1978.

10. Zadeh, L.A.: Outline of a New Approach to the Analysis of Complex Systems and Decision Processes. IEEE Transactions on Systems, Man and Cybernetics, SMC-3 (Januray 1973), pp. 28-44.

11. Bossel, H.: Modelling of the Cognitive Processes Determining Non-routine Behavior. In H. Bossel (ed.): Concepts and Tools of Computer-assisted Policy Analysis. Birkhäuser, Basel, 1977, Ch. 13, pp. 457-481.

12. Müller-Reissmann, K.F., Rechenmann, F.: Cognitive Systems Analysis: An Interactive Program for the Modelling of Deduction. In H. Bossel (ed.): Concepts and Tools of Computer-assisted Policy Analysis. Birkhäuser, Basel, 1977, Ch. 14, pp. 482-537.

13. Bossel, H.: Orientors of Nonroutine Behavior. In H. Bossel (ed.): Concepts and Tools of Computer-assisted Policy Analysis. Birk-häuser, Basel, 1977, Ch. 6, pp. 227-265.

14. Müller-Reissmann, K.F., Bossel, H.: Zur Simulation kognitiver Prozesse bei Entscheidungen: Auf dem Wege zu einer Synthese von Wertforschung und Systemanalyse. In H. Klages, P. Kmieciak (eds.): <u>Wertwandel und gesellschaftlicher Wandel</u>. Campus, München, 1979.

15. Bossel, H., Strobel, M.: Experiments with an "Intelligent" World Model. <u>Futures</u>, June 1978, pp. 191-212.

16. Müller-Reissmann, K.F., Bossel, H.: Kriterien für Energieversorgungssysteme, Institute für Angewandte Systemforschung und Prognose(ISP), Hannover 1979.

17. Bossel, H.: <u>Bürgerinitiativen entwerfen die Zukunft</u>. Fischer, Frankfurt/M., 1978.

# ECONOMIC MODELS OF PERIODIC MARKETING SYSTEMS

A.H. Zemanian

State University of New York, Stony Brook,
N.Y. 11794, U.S.A.

## Abstract

In third-world economies, staple food crops and many other
consumer items are usually marketed through periodic markets. If the
size of a marketing system is measured by the number of its participants or even by the monetary value of the goods exchanged, periodic
markets are commonly the largest component of the overall marketing
system. Our current research is aimed at modelling such systems. A
typical model is a large number of nonlinear difference equations representing the interactions of supply and demand within and between the
various markets in the system. From this we have obtained a number of
qualitative conclusions regarding the temporal and spatial characteristics of price and commodity-flow variations.

## I. Introduction

Periodic markets are common throughout the third world. Indeed, in many developing countries they comprise the dominant kind of
marketing system. Much of the trade in staple food crops at the lowest
levels of the marketing system is performed in them. Although there
are many anthropological, sociological, and economic descriptions and
much empirical data on periodic markets, there are comparatively few
mathematical analyses of them, and those analyses are mainly concerned
with their spatial properties as viewed from central place theory.
Virtually all of the theoretical literature on the spacio-temporal
structure of periodic markets has been concerned with such questions
as the following. How did periodic markets originate? Where are they
located? What is their hierarchal structure? How do their time
schedules synchronize? Which firms are mobile and which are fixed?
What kinds of routes do mobile traders follow? However, there has been
hardly any theorizing about the dynamics of price and commodity-flow
variations over space and time. What is needed are adequate dynamic

economic models of the various kinds of periodic marketing systems.
Moreover, such models will have to be devised if the basic marketing
systems of third-world countries are ever to be incorporated into
global models of third-world economies.

In recent years we have undertaken a research program to
fulfill this need. The present paper is a survey of the results obtained so far. We start by summarizing some typical characteristics
of periodic marketing networks. Next, we describe our model for the
trader. He is in fact the key economic agent, for he is the one that
integrates the spacially separated markets of a developing country into
a single marketing system. Our model of the trader is then used to
construct a model of a periodic marketing system consisting of a single
ring of markets operating on a time-staggered schedule. In a similar
fashion we can construct more general models of a bipartite periodic
marketing network, a marketing system composed of many trader rings
wherein traders buy goods in urban centers and sell them in rural markets, and another such system where traders buy goods in rural markets
for resale in urban wholesale markets. Finally, we mention the modifications that arise when traders are allowed to store goods. The principal conclusions that have been drawn from these models are also
summarized.

## II.  Some Characteristics of Third-world Marketing Systems

The marketing systems of the less developed countries are
considerably different from the marketing systems of the developed
countries. In contrast to the latter, the former commonly have the
following characteristics. (See for example 1, 4 to 13.  A thorough
bibliography on periodic markets is given in 2. and 3.  See also the
Newsletters of the Working Group on Market-Place Exchange Systems,
International Geographical Union for some current literature).

1.  The news about conditions in neighbouring markets may be
poor and for distant markets very meagre if not entirely lacking.  Thus,
agents are usually well-informed about the markets they are trading in
but poorly informed about current conditions in neighbouring markets
and unknowing about distant markets.

2.  Transportation can be quite poor.  As a result, markets
at some distance from one another may be effectively isolated so far
as price arbitraging is concerned.  Goods may pass between those markets through a chain of intermediate markets, but they will then do so
through many hands, many transactions, and much time.

3.  Markets may be held periodically, perhaps every fourth
day as in Nigeria.  Alternatively, they may meet daily, but one or two

days in a period on n days may be major market days, involving bulking, wholesaling, and redistributing for shipments to and from distant markets, with the remaining days being minor market days involving only local trade. This occurs for example in Java where n = 5.

At a more primitive level, traders may pass through a rural producing region only once or twice during a harvest season buying up supplies from the farmers. In this case, farmers will participate in the marketing system only on these occasions and in a haphazard fashion.

4. Periodic markets may form ring clusters such that the markets in a single cluster meet on a staggered schedule with only one or two markets in the cluster open on any particular day. This permits a trader to traverse most or all of the cluster by being in a different market on each day and thereby covering most of the territory serviced by the cluster.

5. Since capital is scarce and costly while labor is plentiful and cheap, there being much unemployment in rural areas of third-world countries, the initial and final transactions of the marketing chain is usually conducted by petty operators working with few stocks, meagre capital, and small profit margins. This means that there tend to be many agents and keen competition at these stages. To a lesser extent the same is true for the intermediate stages of wholesaling, bulking and transportation to distant markets.

6. The responsiveness of the marketing network may be hampered by state interference, such as for example limits placed on the amount of goods traders may transport, or governmental monopolies in the exporting of cash crops. That responsiveness may also be hampered by the lack of credit facilities or the mistrust between different tribes. These factors tend to stunt the development of marketing systems from the two-level form into a more advanced hierarchal form.

As a result of all these characteristics, marketing networks in third-world countries tend to be sluggish. That is, a disturbance in supply or demand at one point of the network will be transmitted throughout the network rather slowly, with distant markets reacting weakly or perhaps not at all. Thus, the impact of the disturbance is absorbed locally without the resources of the whole system being utilized to moderate the local stress (as would happen in the marketing systems of the first-world nations). Consequently, prices are more volatile at local markets but tend to be insensitive to variations in distant markets.

Furthermore, the marketing networks of the third-world countries appear in a greater variety of forms than do the highly integrated·

marketing networks of the first-world countries. The former range from
the bartering practices of primitive societies, to opportunistic trade
in exotic goods with outside traders, to simple producer-to-consumer
local marketing, to two-level periodic and staggered marketing, and
finally to the redistributive hierarchal marketing networks that
approach the systems appearing in the first-world countries. Moreover,
within a single less-developed country, two or more of these systems
may appear simultaneously depending upon the commodity traded (e.g.,
whether it is a locally produced item for local consumption, or an item
produced in one region of the country for consumption in another region,
or perhaps an exported cash crop).

The improvement of marketing being an integral part of the
development of a third-world country, it is important to have a mathematical - as well as a descriptive - understanding of how these various
kinds of marketing systems work. Mathematical analyses of economic
systems are perforce more idealized and usually less realistic than the
nonmathematical investigation of such systems, but they do quantify the
questions at hand, yielding thereby greater precision and a more thorough understanding within the limits of the assumptions necessarily
imposed by the mathematical framework of the study. Nevertheless,
there has been hardly any mathematical-modelling studies of third-world
periodic marketing systems so far as dynamic behavior is concerned.
The aim of our current research is to conduct such an investigation.

The fact that the marketing systems in the third world appear
in a variety of forms is an obstacle toward achieving a comprehensive
mathematical analysis. The situation is quite analogous to that of
economic systems in advanced countries. It is near impossible to find
in actual practice perfect competition in a market, or a pure monopoly,
or an oligopoly satisfying one of the several sets of behavior assumptions made about oligopolies, for example. Yet these ideal systems are
studied nevertheless, for a knowledge of their characteristics leads to
a better understanding of the mixed systems usually found in real-world
situations. We are following the same approach with regard to third-
world marketing systems. Pure forms of the various kinds of periodic
marketing systems are being modelled and then compared to see what
might occur when an actual marketing network lies somewhere in between
the pure forms.

III. The Trader

With regard to price arbitraging between markets, the key
agent is the trader. We assume that he is a rational economic agent,
indeed, a profit maximizing firm that supplies the service of

transferring goods and ownership over space and time. We consider the case where he buys goods in one market for sale in another. By examining his costs of operation and treating him as a profit maximizing firm, one can derive an excess supply curve that explicates his behavior in a given market[16]. This is shown in Figure 1. There, p denotes price and q the quantity of the commodity at hand. The subscript s is the index for the particular market $\phi_s$ in which the trader is operating at time t. $C_s(t)$ is the amount of goods the trader will transport to his next market $\phi_{s+1}$. Thus, $C_{s-1}(t-1)$ is the amount of goods he has brought into the market $\phi_s$. $E_s(t)$ denotes the price he expects to receive in his next market $\phi_{s+1}$, and T is the minimum per unit average variable cost of transferring goods from $\phi_s$ to $\phi_{s+1}$. Thus, if pure competition holds and if the price in $\phi_s$ at t is $P_s(t)$, then the amount of goods the trader sells in $\phi_s$ at t is $Q_s(t)$. If, however, $P_s(t) < P_s^t$, $Q_s(t)$ will be negative; this means that the trader actually buys goods. This reflects the fact that when the difference $E_s(t) - P_s(t)$ is large enough, the trader will be induced to buy goods in $\phi_s$ for sale in $\phi_{s+1}$. The derivation of this behavioral characteristic, which is based on the standard theory of the firm, is given in 16. We now describe how this characteristic can be used to model a particular marketing system.

## IV. A Periodic Marketing Ring

Assume that the commodity at hand is a manufactured item, such as an article of clothing, which is bought by traders in an urban wholesale market and distributed by them to a number of rural markets. We examine the case of an isolated ring of four periodic markets operating on a four-day marketing week. On the first day t = 1, various traders, each described by a characteristic of the form of Figure 1, buy goods in the wholesale market $\phi_1$. They transport those goods to the rural markets $\phi_2$, $\phi_3$, and $\phi_4$ and sell them on days t = 2, t = 3, and t = 4, respectively. This process repeats itself week by week. Assuming that at the beginning of day t = 1 the traders arrive in $\phi_1$ empty-handed (that is, $C_4(0) = 0$) and letting $E_1(1)$ denote their expected price in $\phi_2$ and $T_1$ the minimum per-unit average variable cost of transporting goods between $\phi_1$ and $\phi_2$, we aggregate the resulting excess-supply curves for all the traders in $\phi_1$. Since the excess-supply curves are all negative, we have in fact an aggregate demand curve indicated by the increasing curve of Figure 2(a). On the other hand, the wholesalers in $\phi_1$, from which the traders buy goods, are characterized by an aggregate supply curve, which on the axes of

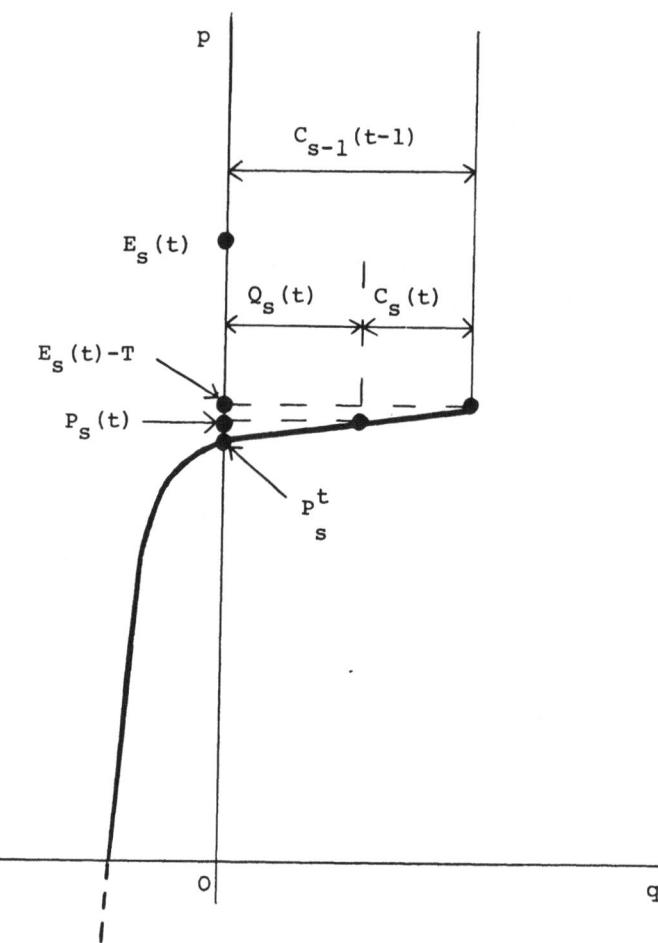

Figure 1.

125

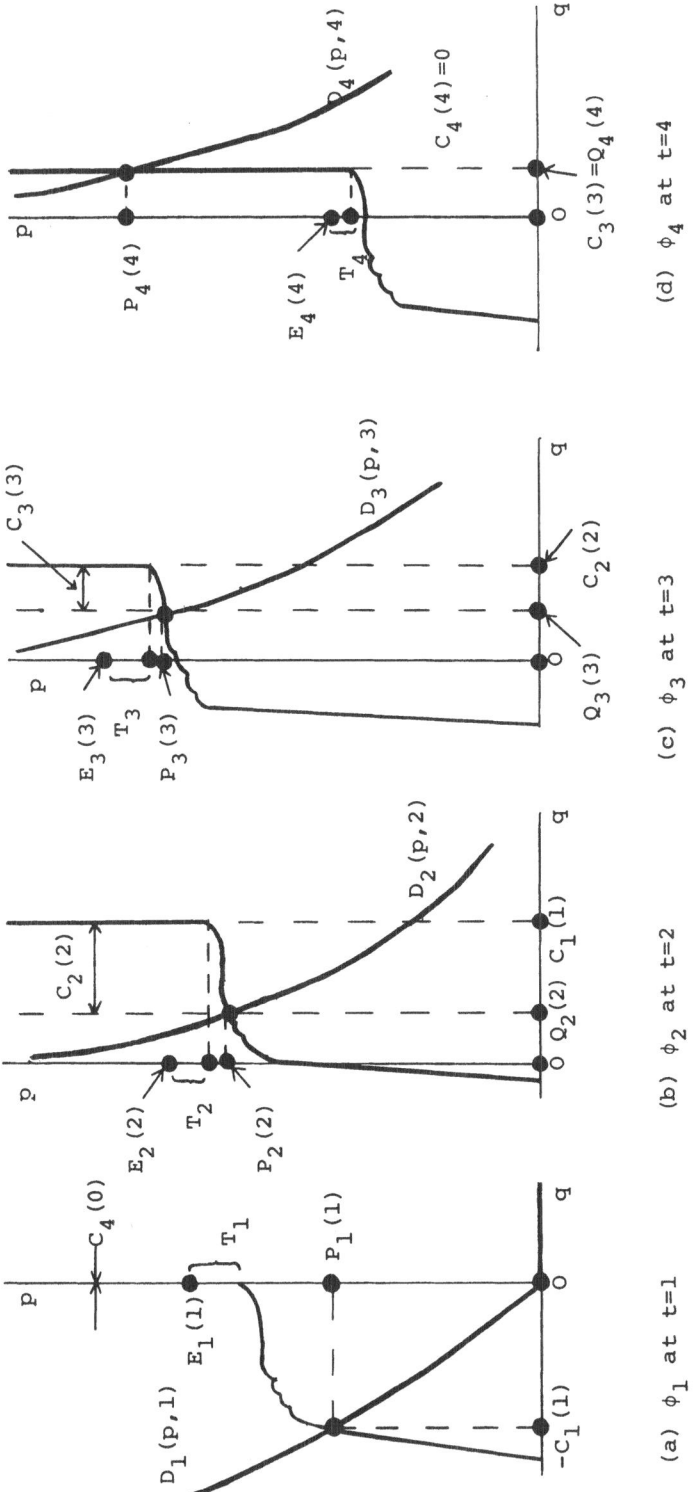

Figure 2.

Figure 2(a) appears as the negative demand curve $D_1(p,1)$. The market $\phi_1$ clears at the price $P_1(1)$ and the quantity $C_1(1)$, the latter being transported to the market $\phi_2$.

To find how $\phi_2$ clears at $t = 2$, we may once again aggregate the excess-supply curves of the individual traders. But, now that aggregate curve is shifted to the right by the amount $C_1(1)$ since this is what the traders bring into $\phi_2$. In Figure 2(b) this yields the rising curve, which is the aggregate excess-supply curve of the traders. Moreover, the traders are faced by an aggregate demand curve $D_2(p,2)$ from the consumers in $\phi_2$, who are primarily local farmers. Thus, the market clears at price $P_2(2)$ and quantity $Q_2(2)$. $C_2(2) = C_1(1)-Q_2(2)$ is the amount that remains in the hands of the traders and is transported to the next market $\phi_3$. The process repeats itself in $\phi_3$ and also $\phi_4$, with each trader selling some or all of his stock and transporting the unsold portion to his next market.

Under normal conditions, prices in $\phi_1$ are comparatively low whereas prices in $\phi_2$, $\phi_3$, and $\phi_4$ are higher. This means that the price $E_4(4)$ the traders expect in $\phi_1$ at $t = 5$ while operating in $\phi_4$ at $t = 4$ is low. The result is that the demand curve $D_4(p,4)$ intersects the excess-supply curve in $\phi_4$ on the latter's strictly vertical portion, which means that the traders sell all their remaining stock in $\phi_4$ and return to $\phi_1$ empty handed.

However, the latter need not happen. Nevertheless, our model will still apply. It holds whether or not traders sell out in any market and whether they are faced by an aggregate demand schedule or aggregate supply schedule from the other agents in their markets. Moreover, it is possible for some traders to buy goods in a particular market while other traders sell them there. The flexibility of our model allows it to be used in a variety of different marketing systems.

## V.   Other Kinds of Periodic Marketing Systems

It is unusual to find an isolated marketing ring like the one described in the preceding section. The common periodic marketing system is a very large one consisting of many intersecting but different marketing rings. Each ring is defined by the traders that follow it. In addition to the local agents in a particular market, a trader will interact with and compete against other traders who come from and go to markets that are not in the first trader's ring. This merely has the effect of complicating the aggregation of the excess-supply curves of the various traders. The resulting equations require a more elaborate indexing system. Nevertheless, the basic ideas, illustrated in

Figure 2, remain the same. In short, the system is described by a very large number of simultaneous nonlinear difference equations (16, Section 4), something that is quite appropriate for computer modelling.

Another kind of model arises when the commodity at hand is bought by traders in a series of rural markets and then shipped by them to urban centers for sale to wholesalers or retailers. This is the common situation for staple food crops. The appropriate model for this case is in a sense the reverse of that described in the preceding section, but there are important dissimilarities. For one thing, it is most unlikely that the traders will carry the commodity back to the rural markets if they fail to sell all of their shipments in the urban centers. It is far more likely that they will store their unsold stock if that commodity is at all storable. This requires a storage schedule for each trader. These too can be aggregated to get an overall analysis of the storage occurring in the system[15].

Special cases of the aforementioned models occur when the marketing system is bipartite, that is, when the traders alternate between one urban center and one rural market. This is the situation that arises more commonly with daily markets, when for example the traders gather a food commodity, such as milk or eggs, in the rural areas during the early morning hours and sell them in urban areas during the day. The resulting models are simplifications of the aforemention-ed ones and are presented in detail in (14), and (15).

## VI. Qualitative Results

Since the models we have obtained require the aggregate excess-supply curves of many different groups of traders, it will be virtually impossible to use them to obtain a quantitative description of actual periodic marketing networks. However, they can be used to derive qualitative properties of those systems.

For example, once a dynamic model becomes available, a natural question to ask is whether it has an equilibrium state. This question does not appear to have been previously asked about periodic marketing systems evidently because dynamic models had not been avail-able for them. The importance of this issue lies in the following facts. If it can be shown that no equilibrium state exists, then we can conclude that the periodic marketing system is condemned to per-petual price variations even when the exogenous supply and demand func-tions are fixed with respect to time. One might then ask what regu-larities whose variations might have. For example, do limit cycles exist, and, if so, are they stable? On the other hand, if an equilibrium state exists, then we have the possibility of steady and predictable prices, upon which rational economic planning can be based. In the latter case, the questions of uniqueness and stability for the equilibrium state can be investigated.

For our models of bipartite periodic marketing networks, we have been able to establish the existence of equilibrium states, and, in the case where no storage is allowed, we have shown that the equilibrium state is unique and asymptotically stable. When storage is allowed, the uniqueness of the equilibrium state and its asymptotic stability remain open questions. For a periodic marketing system consisting of many rings, the question of the existence of an equilibrium state is still open. We have been able to establish the existence (and uniqueness, as well) of an equilibrium state only in the special case where the marketing system consists of a single ring.

A phenomenon peculiar to periodic marketing systems is the step-by-step transmission of price disturbances, a characteristic identified and described by W.O. Jones (8, p. 119). This phenomenon is reproduced by our dynamic models. Moreover, our models permit us to examine the resulting price signals and the relative sizes of price oscillations in representative computer models of periodic marketing systems.

Another category of problems concerns the occasionally surprising behavior of periodic markets (8, pp 24-25); in fact, price variations appear at times to be erratic and unpredictable (10, pp 21-22). Our dynamic models provide possible explanations for these phenomena. In 15, section 9 we have shown how a sudden oversupply in one rural market of a bipartite periodic marketing network can generate an initial price signal for a shortfall propagating in one direction of the network and in initial price signal for an oversupply propagating in another direction. This is caused by the "cutoff" phenomenon, where all the traders on one or more legs of the marketing network stop trading because of unfavorable prices at their markets. In 16, section 5 we have shown that, for a ring-type periodic marketing system, a sudden shortfall in one urban center can lead to a price rise (rather than a price fall) in a rural market two market days later. Moreover, this latter phenomenon occurs in the abscence of cutoff.

In addition to these results, the most important contribution of our studies of periodic marketing systems is we feel a methodological one. It provides for the first time a means of performing computer studies of the dynamic economic behavior of the kind of marketing system that is ubiquitous throughout the third world.

## References

1. Isaac A. Adalemo, "Traders' Travel Patterns, Marketing Rings and Patterns of Market Shift," _Nigerian Geographical Journal_, Vol. 18 (1975), pp. 17-26.

2. R.J. Bromley, _Periodic Markets, Daily Markets and Fairs: A Bibliography_, Centre for Development Studies, University College of Swansea, Great Britain, 1974.

3. R.J. Bromley, _Periodic Markets, Daily Markets and Fairs: A Bibliography Supplement to 1979_, Centre for Development Studies, University College of Swansea, Great Britain, 1979.

4. R.J. Bromley, "Trader Mobility in Systems of Periodic and Daily Markets," in D.T. Herbert and R.J. Johnston (Editors) _Geography and the Urban Environment_, Vol. III, John Wiley, New York, 1980.

5. R.J. Bromley, Richard Symanski, and Charles M. Good, "The Rationale of Periodic Markets," _Annals of the Association of American Geographers_, Vol. 65 (1975), pp. 530-537.

6. Alan M. Hay and Keith S.O. Beavon, "Periodic Marketing: A Preliminary Graphical Analysis of the Conditions for Part-Time and Mobile Marketing," _Tijdschrift voor Economische en Sociale Geografie_, Vol. 71 (1979), pp. 27-34.

7. Donald W. Jones, "Production, Consumption, and the Allocation of Labor by a Peasant in a Periodic Marketing System," _Geographical Analysis_, Vol. 10 (1978), pp. 13-30.

8. W.O. Jones, "The Structure of Staple Food Marketing in Nigeria as as Revealed by Price Analysis," _Food Research Institute Studies_, Stanford University, Vol. 8 (1968), pp. 95-123.

9. W.O. Jones, _Marketing Staple Food Crops in Tropical Africa_, Cornell University Press, Ithaca, N.Y., 1972.

10. W.O. Jones, "Regional Analysis and Agricultural Marketing Research in Tropical Africa: Concepts and Experience," _Food Research Institute Studies_, Vol. 13 (1974), pp. 3-28.

11. Robert H.T. Smith, "Periodic Market Places and Periodic Marketing: Review and Prospects," _Progress in Human Geography_, to appear.

12. Richard Symanski and M.J. Weber, "Complex Periodic Market Cycles," _Annals of the Association of American Geographers_, Vol. 64 (1974), pp. 203-213.

13. Keith J. Tinkler, "The Topology of Rural Periodic Market Systems," _Geografiska Annaler_, Vol. 55B (1973), pp. 121-133.

14. A. H. Zemanian, "Two-level Periodic Marketing Networks Without Market News," _J. Math. Anal. Appl._, Vol. 68 (1979), pp. 509-525.

15. A.H. Zemanian, "Two-level Periodic Marketing Networks Wherein Traders Store Goods," _Geographical Analysis_, to appear.

16. A.H. Zemanian, _A Dynamic Economic Model of Periodic Marketing Rings_, State University of New York at Stony Brook, College of Engineering Tch. Rep. No. 340, April 16, 1980.

# A NEW URBAN TRAVEL MODEL

Yacov Zahavi

Mobility Systems, Inc. USA

## 1. Introduction

Urban travel models are perhaps not the best example of how to structure a global model. Nonetheless, a city can be viewed as a system not less complex than a group of settlements in a country, or a group of nations in the world. This is especially true when the concepts underlying such models are examined closely. When decomposing the problems to be modeled, human behavior under constraints appears to be a common denominator in most of them. Thus, it is hoped that this paper will add its share to a better understanding of human behavior.

This paper presents intermediate results from an on-going effort to develop a new urban travel model. This model, called the Unified Mechanism of Travel, or UMOT for short, was first conceptualized for the World Bank, and further developed for the U.S. Department of Transportation (Zahavi, 1979). It is being extended now to include urban structure within a dynamic feedback framework.

A principal reason for the attempt to develop the UMOT model was dissatisfaction with the operation of conventional urban travel models under rapid and fluctuating changes in their inputs, such as fuel prices. However, in order to develop a better model, one has first to identify what might be wrong with previous models. While doing so, it became increasingly evident that conventional urban travel models displayed many conceptual difficulties, most of which seemed to be in conflict with basic principles known in system theory (Kalman, 1978).

Since the author´s discipline is in the field of travel models, not global models, the following list of conceptual difficulties applies specifically to conventional urban travel models. Even so, it will come as no surprise if some of them may seem to be familiar to modelers in other fields.

The paper is divided into two principal parts. The first part details some of the basic problems encountered in conventional urban travel models, as well as the way in which they are resolved in the UMOT model. The second part presents the general structure of the UMOT travel model, together with several reflections on its extension to include urban structure. The purpose of these two parts is to encourage, and benefit from, a dialogue between modelers of different disciplines about their experience in modeling large scale systems of human interactions.

Because of the complexity of the subjects discussed in this paper, and in light of the space constraint imposed on it, the paper is presented in a simple, telegraphic, form.

## 2. The UMOT Model vs. Conventional Travel Models

The UMOT model is based on the predictable regularities observed in the mean travel time and money expenditures per representative traveler of different socio-economic groups. Since these regularities are observed to be transferable both between cities and over time in a country, the expenditures are regarded as "travel budgets" which, under certain conditions, are applied as constraints on travel behavior.

One useful way of applying travel budgets as constraints is within the microeconomic theory of consumer behavior, where consumer utilities are maximized under explicit constraints. By this theory, as one of several available to the UMOT model, the utility of spatial and economic opportunities to which a person travels, represented by the average daily travel distance, is maximized under the explicit constraints of time and money budgets allocated to travel.

The application of explicit constraints is a powerful tool, since the constraints eliminate the need for much of the coefficient calibration of conventional models. Thus, once the constraints and the unit costs of all alternative models are known, the model produces estimates of such travel characteristics as daily travel distance, modal share, and car ownership.

Perhaps the best way of explaining the UMOT´s distinctive aspects is by listing a few of its unique characteristics, which distinguish it from conventional travel models.

Causality - Causality in modeling travel behavior is typically assumed *a priori*. For example, it is typically assumed in travel models that car availability per household increases trip generation. Namely, car ownership is the cause, while more trips is the effect. However, it might be also argued legitimately that the need for more travel generates car ownership levels.

In the UMOT model there are no assumptions about unilateral, fixed, causality. The process is based on a systemwise approach, where all travel components interact with each other and with the transport system through a simultaneous dynamic feedback process. Thus, each component can be both cause and effect, depending on the feedback step.

Calibration and Validation - Conventional travel models are calibrated to the observed travel choices. Thus, both the independent and dependent variables must be known before such models can be calibrated. For instance, a model which is required to estimate trip rates per household is calibrated (fitted) to the observed trip rates, and the calibration process becomes a balancing act between the two sides of an equation. Such a model is then validated by its ability to reproduce the same observations to which it was fitted. This may be regarded as a tautological process.

In the UMOT model no desired output is ever calibrated to the observed values. The outputs are the expected choices, which are then compared with the observed choices - not fitted to them - for the model's validation.

For example, the process can be started by assuming that each and every household in the urban area owns, say, 5 cars. Such an assumption, of course, is absurd. Nonetheless, the travel system converges rapidly to the observed car ownership levels, by the households' socio-economic characteristics.

Transferability - Conventional travel models usually must be recalibrated in each separate city. The coefficients, fitted to cross-sectional data, are then assumed to remain fixed over time for each city. However, a prerequisite for a model's temporal transferability in one city is considered to be its spatial transferability between cities at one point in time, a condition which is not always met by conventional models.

The UMOT model is based on the relationships that apply to the travel budgets, relationships which have been observed to be transferable both spatially and temporally in one country. Furthermore, there are no fixed coefficients associated with the choices in the UMOT process, and the model is activated through all its phases for each endogenous and exogenous change. Even the constraints are not constant, but can vary in response to endogenous and exogenous factors.

Stable Parameters - Conventional models are based on trips. However, the definition of a trip, and its related trip distance, trip time and trip cost, is ambiguous to a large extent since trips are linked/chained/clustered and combined into "tours" in various ways

during the calibration phase of the models. Thus, the basic travel
data in a city can vary according to the chosen definition of a trip,
resulting in different models.

The UMOT model is based on travel components that remain unchanged by any definition, which are the <u>total daily</u> travel components,
such as the daily travel distance, and the daily travel time and money
expenditures per traveler/household. Thus, there can be only one model
for one data set.

<u>Organization Process</u> - A conventional travel model is divided
into many sub-models; car ownership sub-model, trip-generation sub-
model, trip mode-choice sub-model, and trip-distribution sub-model.
Each sub-model is further sub-divided by trip purpose and/or mode, with
no explicit interactions between the various sub-models. Moreover,
each sub-model is based on a large number of independent variables,
and one disaggregate car-ownership sub-model boasts of 21 variables
which are used in various combinations for different population segments. An additional problem is that many of the independent variables are used repeatedly and/or alternately in separate sub-models.
For example, income is used as one independent variable of several in
the car ownership sub-model, while later car ownership <u>and</u> income are
used as two of the independent variables in the trip generation sub-
model, and so on. Although multicolinearity is recognized, the multi-
plicity of variables is kept in such models, mostly in order to allow
each explanatory variable to be forecasted separately in each sub-model.

In the UMOT model there is only one organizing principle
which operates the travel system: an objective function which represents the individual travelers/households attempts to maximize their
spatial and economic opportunities within their travel constraints,
with dynamic feedback between the individual travelers/households and
the transport system. Furthermore, factors which are regarded in conventional models as independent or dependent variables of the so-called
causal relationships, and which often alternate between sub-models,
are regarded in the UMOT model as either input to, or output from, the
process. For instance, income is an explicit input to, while car ownership is an explicit output from, the process. In short, in the UMOT
model there is only one organizing principle, which encompasses all
population segments and all travel components within one interactive
process.

<u>Equilibrium vs. Disequilibrium</u> - Conventional urban travel
models are usually based on the assumption that the demand is in equilibrium with the supply. Thus, by definition, each alternative
scenario must reach, or at least approach, equilibrium between demand

and supply.

However, it is equally valid to say that it is the amount of possible disequilibrium associated with alternative futures which generates forces that dynamically change urban structure, often in unexpected ways. The UMOT model attempts to measure, as one of its outputs, the amount of potential disequilibrium affecting households of different socio-economic and locational characteristics.

As can be seen from the above examples, there are some significant differences between the UMOT travel model and conventional travel models. Such differences are reflected not only in the concepts underlying the models, but also in the way the models are activated. To mention one example: conventional models regard travel distance as a disutility, measured by the time and money costs required to overcome distance between origin-destination pairs. In the UMOT model daily travel distance is regarded in utility terms, representing the benefits of access to spatial and economic opportunities within the urban area. Such a conceptual difference literally turns the conventional modeling process upside-down; daily travel distance is a final output of conventional models, while it is the first output of the UMOT process.

Two comments are now called for, in order to put the above comparisons in the right persepctive. First, putting all conventional travel models under one heading probably does injustice to some of them. The purpose of the comparisons, however, is to emphasize the prevailing general approach to modeling travel, rather than to single out a specific model. The second comment is that the UMOT model is not yet an operational model. It is still in its development stages, and although results have been very encouraging, only further development will tell whether it is successful in fulfilling its expectations.

A simplified description of the UMOT structure is presented in the following section, in order to enable the reader to judge better the systemwise approach underlying the UMOT travel model, and perhaps also assist in its further development.

## 3. The UMOT Structure

Figure 1 presents a simplified flow chart of the UMOT travel model, showing the feedback processes.

Perhaps the best way of describing the systemwise approach used in the UMOT model is by an example. Both travel time and money budgets affect, and are affected by, the level of car ownership of a specific household. The process can be started, for example, by assuming that each and every household in the urban area owns, say, 5 cars.

135

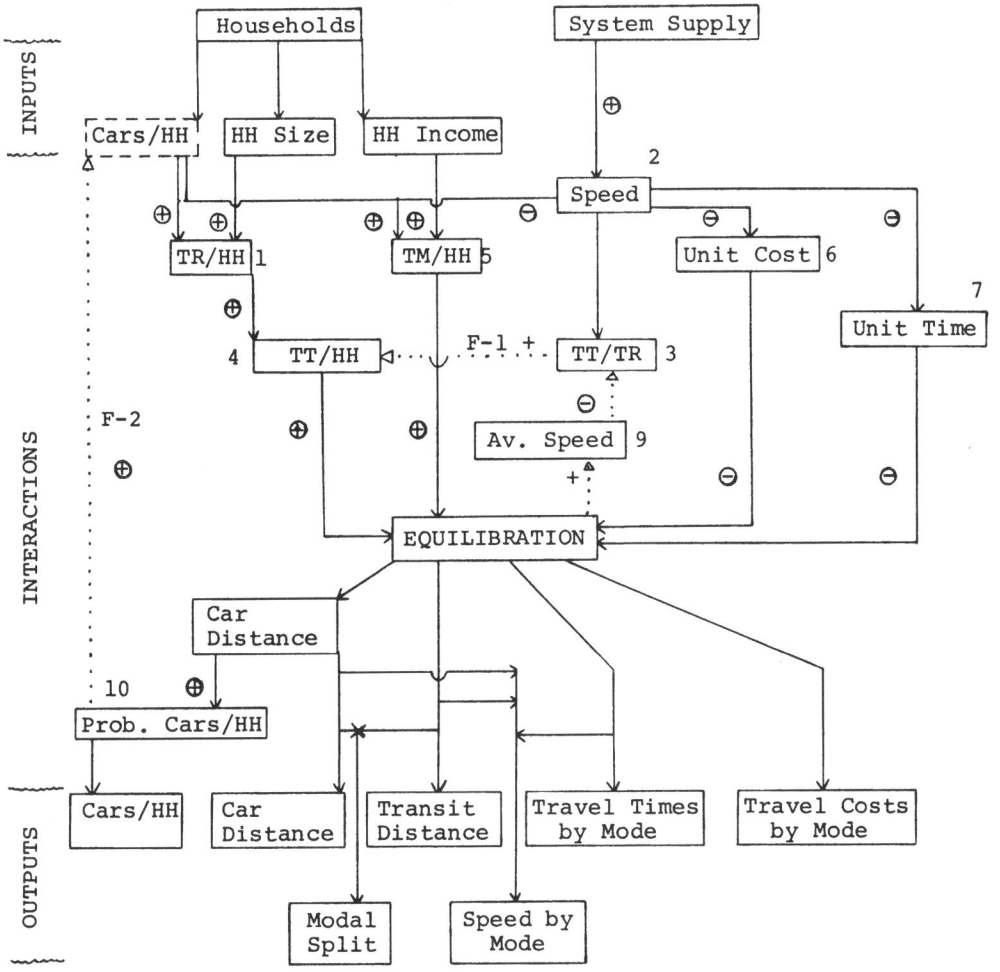

Figure 1.  Flow Chart of the Interactions between Travel Demand,
System Supply and Car Ownership, the UMOT Model

———→ Input/output flow
    5 Interaction.  Effect of Input on Output is expressed by  + or -
· · · · · ▷ Feedback

Such an assumption, of course, is absurd, but at this stage the travel constraints take over and drive the travel system, based on given unit costs, to result in (i) the estimated travel distance by each mode, (ii) the estimated car ownership levels, (iii) the interaction between the estimated number of cars and a given road network results in new unit costs of travel, which are fed back into the travel distance phase, and (iv) repeating the process by iterations results in the rapid convergence of the travel system (where all travel components, including the travel budgets, interact with each other) to outputs which agree with the observed travel and car ownership levels in the urban area.

In short, there is no need to calibrate separate sub-models for the travel components, such as for car ownership levels; the process responds to any input, even an absurd one, and it adjusts the system to converge to expected values within the organization principles. Borrowing an expression, the UMOT process can be termed as a "self-organizing system". In all cases tested until now, the expected values of travel components were found to match the observed ones.

At this stage one more point should be brought up. It was mentioned in Section 2 that predictable regularities had been observed in travel time and money expenditures per average traveler/household. This does not mean that each and every traveler/household spends a fixed amount of time and money on travel each day. Predictable regularities were also observed in the variations around the mean values. For instance, the coefficient of variation (standard deviation over mean) of travel time per average traveler was found to be similar in a wide range of cities in both developing and developed countries, and for different population segmentations. Thus, once the mean and the coefficient of variation per average traveler belonging to a population segment are known, the probability of an individual traveler to behave as his group can be deduced.

The last subject to be mentioned in this paper is the integration of travel and urban structure within the UMOT process, a project which is presently in progress for the U.S. Department of Transportation (Mobility Systems, 1980). The theoretical part is an extension of the UMOT travel utility theory to urban structure, including residence-job locations and the dynamic effects of changes in endogenous factors (e.g., household income, household size) and exogenous factors (e.g., transport system, travel costs) on urban structure. The empirical part is the study of travel probability fields, as explained below. It was encouraging to find out that the theory predicted the observations. Because of lack of space, only the travel probability fields are mentioned, below.

Conventional travel models deal with the spatial distribution of trips through the trip distribution sub-model, which requires lengthy calibrations of trip origin-destination matrices and travel impedance relationships, as well as an extensive description of the transport system. The notion of travel probability fields, on the other hand, is based on the predictable regularities observed in travel time and money expenditures and their extension to the spatial distribution of trips. Instead of dealing with single trips, it is possible to consider, and describe, them in terms of a continuous statistical distribution. Ineeed, it was verified that the travel probability fields can be described statistically, and that they respond to endogenous and exogenous factors in expected ways. Such travel probability fields are now being developed as a direct link in the representation of urban structure as a variable in the UMOT urban model.

An additional subject which is now under investigation is the application of bifurcation and catastrophe theories in the UMOT model, in order to describe possible sudden changes (jumps) in the travel behavior and/or urban structure under continuous changes in exogenous factors, such as travel costs. Conventional urban travel and urban structure models, which are continuous in form, are unable to identify and describe critical points after which there is a sudden change in the behavior of individuals, or a city, under continuously increasing pressures, although such phenomena are known to occur. But this, of course, is a subject for an additional paper.

ACKNOWLEDGMENTS

The author is grateful to R.W. Crosby and D. Kahn, of the U.S. Department of Transportation, Research and Special Programs Administration, for their permission to refer in this paper to some results from an on-going study. Special appreciation is also extended to R.F. Drenick for his encouragement to write this paper. However, the responsibility for the views expressed in this paper rests solely with the author.

## References

Kalman, R.E. (1978). "A System-Theoretic Critique of Dynamic Economic Models". <u>Global and Large Scale System Models</u>. Proceedings of the Center for Advanced Studies (CAS), International Summer Seminar, Dubrovnik, Yugoslavia. Springer-Verlag, Berlin, Heidelberg, New York, 1979.

Mobility Systems, Inc. (1980). The UMOT/Urban Interactions. Under preparation for the U.S. Department of Transportation, Washington, D.C.

Zahavi, Y. (1979). The UMOT Project. Report DOT-RSPA-DPB-20-3, U.S. Department of Transportation, Washington, D.C., August 1979.

# MODELLING THE SELF-ORGANIZATION OF HUMAN SYSTEMS

P.M. Allen

Chimie-Physique II, C.P. 231, Campus Plaine,
Université Libre de Bruxelles, Brussels 1050, Belgium

## Introduction

The basic truth about policy making and decision taking in
complex systems such as cities, businesses, and even cooking, is that
obtaining harmonious results for evolution remains something of both
an art and a science. And yet, how can this be so? How can it be that
a science capable of the Apollo Project cannot "crack" such problems as
the prediction of urban change, next year´s top selling automobile or
next month´s money supply? The thesis that I shall advance here is
that despite its impressive references the great achievements of science
have pertained to certain classes of system subject to two highly restrictive conditions. Either they deal with "simple" systems which involve only a few interacting particles, where the dynamical trajectory
of each can be strictly followed, and is generally subject to some conservation law, or, if dealing with a complex system of many interacting
elements, then such successful laws as for example those of hydrodynamics, or of chemical kinetics, have only been derived for systems at
or near thermodynamic equilibrium. Here, the system is, microscopic-
ally, in the "most probable" state, and all flows and forces can be
linearized.

When we are interested in finding a reduced description, or
model, of a complex system such as a city, a nation or a firm for example, then neither of these conditions holds. Thus analogies with
planetary motion or thermodymic equilibrium laws have no reason to be
useful or correct, and yet this has been the basis for most systems
modelling. The basic contention being that if we build an interaction
diagram in terms of aggregate variables, then provided we capture the
flow mechanisms of the system correctly, then the evolution of the
system will be modelled correctly. It is assumed that by the "law of
large numbers", if the aggregates are made over sufficiently numerous
actors or elements, then the averages must be a satisfactory representation of this part of the system, and that therefore predictions can

be made from the system of interacting dynamic equations of these aggregate variables. That is, we define dynamical laws for the macro-variables, and we study their evolution by following the system trajectory under these laws. Such an approach is, I believe, not only inadequate for describing human systems, but has been shown to be inadequate for physico-chemical systems.

In fact a new paradigm has emerged over recent years from the study of relatively simple, non-living systems which are, however, composed of many particles, that is they are of macroscopic size $10^{23}$ and are maintained or driven far from thermodynamic equilibrium. This new paradigm is that of "dissipative structure", wherin such systems, providing that they contain sufficient non-linearity in their interactions between the elements, can exhibit a spontaneous process of "self-organization", where structure emerges from uniformity and where symmetries can be broken. During such an evolution, as we shall see, the "law of large numbers", often invoked unconsciously in our thinking, breaks down. That is to say the evolution of such a system is no longer determined completely by the average values or aggregate variables constituting the model. The macroscopic or reduced description proves to be inadequate, and its associated determinism goes out of the window!

Such a paradigm, emerging from the physical sciences, offers us a new basis for modelling complex human systems, and it is not by accident, because such systems are also subject to strong flows of matter and energy which keep them far from thermodynamic equilibrium, and also exhibit strong non-linearities in the cooperative and competitive behaviours of their constituent individuals. Let us therefore devote some space to the understanding of these new concepts associated with self-organization in the well understood area of chemical systems. After that we can use the lessons learned in order to build models or reduced descriptions of complex systems involving humans.

## The Evolutionary Paradigm of Dissipative Structures[1]

In physics there are three basic levels of description. First, there are the classical or quantum laws governing the motion of each particle, presumably constituting the "basic" or "complete" description. Secondly, we have the probabilistic approach which supposes some "unknown" underlying motion of the particles, which is supposed to give rise, for example, to a "birth" and "death" equation governing the evolution of the system. The third level of description is the macroscopic one of thermodynamics, hydrodynamics, chemical kinetics, etc., where the equations governing the evolution of the system are written in terms of variables which are themselves "aggregate" or "average" quantities. This is the reduced description. When is the passage between these levels entirely clear and well understood? This is only satisfactory for systems which are at, or very near to, thermodynamic equilibrium. Only then can the macroscopic, reduced, description be deduced from the "complete" one. However, far from thermodynamic equilibrium description breaks down, together with its associated determinism, and the idea that we can understand and predict the evolution of the system simply in terms of the average values of certain quantities proves to be inadequate.

This breakdown is associated with the occurrence of bifurcations in the solutions of the macroscopic equations. Bifurcations introduce an unsuspected wealth of new phenomena into the otherwise rather trivial evolution of the system, resolving the apparent contradiction in the traditional meanings attached to the word "evolution" in the physical and the human sciences. In the former it has referred to the movement towards thermodynamic equilibrium, the elimination of non-uniformities and the increase of internal disorder, while in sociology and biology it has been associated with increasing complexity, specialization and organization.

In order to understand some of the basic points which arise from these new concepts, let us look briefly at a simple, but remarkable chemical experiment which demonstrates these striking properties. It consists of the relatively simple chemical scheme:

$$A \longrightarrow X$$
$$B + X \longrightarrow Y + D$$
$$2X + Y \longrightarrow 3X$$
$$X \longrightarrow E$$

where, starting from A and B, X is produced, which produces Y, which in turn helps to produce X, and the final products of the reaction are D and E. If "systems" were minded, then in modelling this problem we would draw the very simple interaction scheme shown in Figure 1.

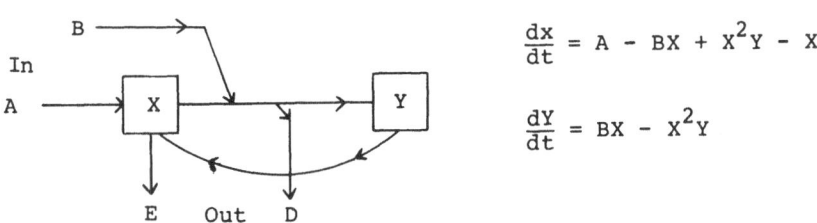

$$\frac{dx}{dt} = A - BX + X^2Y - X$$

$$\frac{dy}{dt} = BX - X^2Y$$

Figure 1. The schematic diagram of interaction of the Brusselator.

Figure 2. An example of dissipative structure.  In the Belousov-
Zhabotinski reaction malonic acid is oxidised by bromate in
the presence of cerium.  When the reaction is performed in
a shallow dish spiral waves develop.

This particular reaction scheme has been studied in detail by the Brussels´ school (it is even known as the Brusselator) and has been found to display various types of self-organizing behaviour. Thus, starting from the uniform, well-mixed, state of the system at equilibrium when the flows in and out of the system of A, B, D and E are zero, we can drive the reactions away from equilibrium by gradually increasing the rate at which A and B are pumped in, and D and E are extracted. At a certain critical distance from equilibrium an instability occurs. This threshold marks the point at which the least fluctuation can cause the system to leave its uniform, well-mixed (maximum disorder) state, and move to some qualitatively different new state of organization. This can be perhaps a stationary pattern of the concentration of X and Y, or moving waves of concentration, even as complex and beautiful as expanding spiral waves. (See Figure 2) Such patterns represent the coherent behaviour of billions and billions of molecules, organized over distances which are absolutely vast compared to that of the molecules. Where does the "information" for such organization reside? What are the necessary conditions that must be met in order to observe such phenomena? The intensive study of these structures, called "dissipative structures", has revealed the answers to these questions over recent years, and gives us a fascinating new perspective on structure and regularities observed in the world which, I believe, is of fundamental importance.

Firstly then, what occurs in the system when a dissipative structure appears? At such a moment, what we witness is the instability of the previous macroscopic pattern of organization. Thus, for example, if we are moving away from equilibrium, we have initially a uniform distribution of the densities of X and Y. Each point in the system is doing the same thing, and because of this there are no internal flows of X and Y between different regions of the system, since there are no strong differences of concentration. However, as reaction rates are increased, the kinetic equations, the model, the reduced description, become ambiguous. That is they permit potentially more than one solution. Thus, in the patterns shown in Figure 2. we see the realization of other macroscopic organizations of the system, where high and low levels of production of X and Y in different regions of the system are balanced by flows of X and Y between them. A new structure such as this occurs then when the old state becomes unstable. That is, when fluctuations around the average values of the variables (remember that the kinetic equations are merely written in terms of averages, a reduced description), which are of course always present, will carry the system off to one of the new states of organization. Which one it is, depends on the precise nature of the fluctuation, which is of course not controlled by the kinetic equations which only discuss, deterministically, the evolution due to the average values of the variables.

Dissipative structures, therefore, invoke both chance and necessity. The conditions required in order to observe them in a system, are that there should be simultaneously more than one solution to the kinetic equations possible, and this in turn requires non-linearity in the interaction terms. (Linearity means only one solution, quadratic dependence two, cubic three, etc.) Thus, if our reduced description involves kinetic equations of change of the variables which are non-linear, then we may expect such a system to have bifurcations in its solutions at certain times and for the fluctuations to play an important role in the evolution of the system, in choosing which of the possible branches of solution the system will in fact take. Complex systems involving feedback will in general give rise to a whole series of bifurcations, as illustrated in Figure 3, and the understanding of its evolution necessarily requires therefore the study of its passage, not just along the particular branch where it happens to be at the initial time, but the possibility of structural reorganizations corresponding to its passage through bifurcation points, and on to new branches, which are perhaps qualitatively different in character.

Such a point of view introduces several important points which I believe are particularly significant for the social sciences. Firstly, it introduces "history" into the explanation of structure, for if we look at Figure 3 we see that, for example, the fact that the system is organized in the manner corresponding to branch C, necessarily implies that in a system growing from initial simplicity, it happened to take the B fork, and before that, fork A. No "explanation" can ever deduce from the reduced description of the problem the unique necessity of finding the system in state C, for parameter value p, because of the ambiguity of the macroscopic description at this point. Secondly, even a system as simple as that shown schematically in Figure 1 has developed a certain autonomy. It is the system not the experimenter who chooses which branch of bifurcation it will take, since this depends explicitly on the particularities of the fluctuation that happens to occur at the moment of instability, and these fluctuations are "outside" the model. Thus, in addition to the fact that, of course, the external environment of a system may change according to a scenario in order to allow for the non-closure of the model to the outside world, we are also obliged to admit that a non-linear dynamical system is not closed either with respect to its interior. This is another way of saying that the world

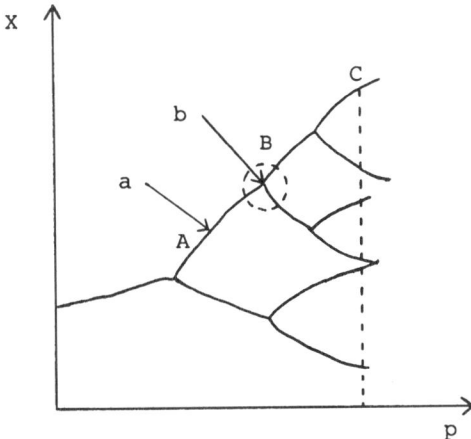

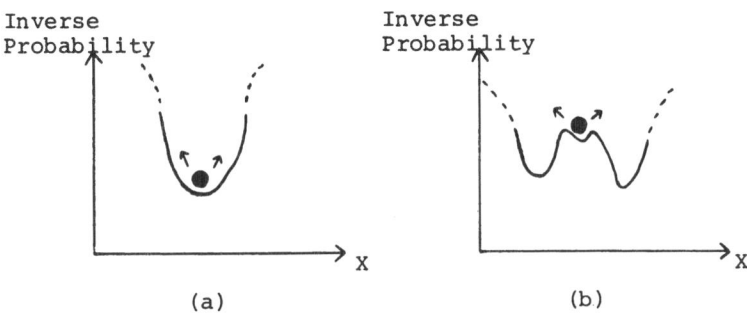

(a)                              (b)

Figure 3. A general bifurcation diagram for a complex system.  Far
         from bifurcation points (a), we have stability, in that small
         fluctuations are damped.  Near a bifurcation point, however,
         (b) it is the fluctuations which decide which branch will be
         taken.  The explanation of the structure corresponding to
         branch C necessarily involves the historical choices of A
         and B.

is always richer than the model and that, at times, some of this richness of the microscopic level breaks through to the macroscopic level
causing a re-organization.  The third highly significant aspect of
dissipative structures is that they already pose the problem of the
"chicken and the egg", related to structure and function.  If we ask
why, we find a particular band of high concentration of X, say, in one
of the examples of Figure 2, it is that its non-linear rate of production (as a function of X and Y) is exactly balanced by the diffusion
of X into the surrounding area with little X.  In other words, if we

now ask why there are strong micro flows of X between different zones, then the answer will be that they are just what are "required" to maintain the production of X. What all this is simply saying is that the kinetic equations have a stable solution of this type. We can now perhaps return to one of the questions that arose earlier. Where is the "information" necessary to organize the system in this way? The answer would seem to be that "information" is probably a misleading word, since it implies somehow that some-one wanted to organize the system. The fact is simply that stable structured solutions are compatible with the kinetic equations of the system, and that the particular structure observed depends firstly on this history of the system and secondly, of course, on the details of the non-linearities, that is the exact values of the parameters of interaction and diffusion. Small changes in these parameters can therefore obviously provoke major reorganizations of the macrostructure, changes which cannot either be viewed as moving in any clear direction, for example, towards an optimization of anything in particular. Furthermore, the Brusselator does not have a potential function governing its evolution, and catastrophe theory is therefore incapable of describing the changes that occur. This is an important point for us, whose primary interest lies with human systems, because a potential function for such a system corresponds to a "global utility function" for a city say, and it gives a rationality to the evolution of a complex system which I do not believe it has. One of the essential elements of the point of view I wish to develop is that a complex system undergoes an "open" evolution, with new properties and new values emerging along an expanding tree of possibilities. This is quite a different view from that in which the system is governed by a potential, albeit one of an interesting shape, where in a sense, everything that can happen is already contained in its specification.

The basic idea I wish to develop, therefore, is that just as the Brusselator can give rise to highly complex structures arising nevertheless from a very simple, but non-linear reaction scheme (Fig. 1.), so perhaps the apparent complexity of human systems may be partly understandable in terms of a few non-linear interaction mechanisms. Thus, by supposing some simple form for the interactions between the actors of a system, we may produce, spontaneously, during the dynamic evolution of the system, a macroscopic self-organization into perhaps a highly complex structure, where structure and function will be enmeshed in the system, recording the particular course of history. Such a non-linear dynamic system is a collective memory.

## A Simple City System

Having tried the readers' patience thus far with perhaps too much physics and philosophy, let me attempt to illustrate in concrete terms what the speculations of the previous section may be able to do for us when faced with the problem of modelling, for example, a city. The aim here is not to describe a completely realistic model, but rather to set out the basic framework, a match-stick drawing as it were, of the "workings" of a city, in the hope of being able to explore the long-term evolution, involving structural changes. We hope from this to be able to build a model which at least can predict the sort of structure that may evolve under a certain scenario, with the accent on the quali-tative features of that structure rather than on quantitative accuracy.

The first step in the operation is then to choose the signi-ficant actors of the system, whose decisions, and the interplay of these, will cause the urban system to evolve. In agreement with much previous work, particularly for example the philosophy of a Lowry-type model, we first consider the basic sector of employment for the city, and in particular two radically different components of this; the industrial base and business and financial employment. Next, we con-sider the service employment generated by the population of the city, and by the basic sectors, supposing two levels, a short-range set of functions and a long-range set. The residents of the city, depending on their type of employment, etc., will exhibit a range of socio-economic behaviour, and for this we have supposed two populations corresponding essentially to "blue" and "white" collar workers.

The next phase of the modelling is in attempting to construct the interaction mechanisms of these variables, which in essence re-quires a knowledge of the values and preferences of the different types of actors represented by the variables and, of course, how these values conflict and reinforce each other as the system evolves. Let us first discuss the mathematical representation of values and prefer-ences in order to set up our system of equations (our kinetic equations which are our reduced description of reality).

The basic step, about which almost all multicriteria analysts agree, is to suppose that each actor can, when faced with a choice, at least list his main criteria of decision: price, facility, prestige, time involved, etc. Let us assume that these factors define the "rationality" of that particular actor, without asking whether or not there is an objective definition of rationality. The second step in attempting to characterize the preference for one or another choice is to assign some appropriate "weighting" to the various criteria already retained so that in some way their relative importance can be taken into account. Let us stress here that we are discussing the preferences of a single actor. Let us look at four possible choices with three dimensions of preferences.

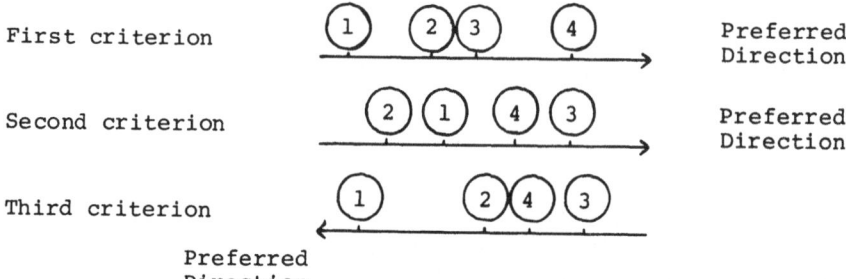

Figure 4. The ordering of four possible actions on axes corresponding to three criteria.

Clearly, if the three criteria were strictly quantitative, were numbers, then it would be possible to simply "add" their weighted values for each choice and identify the "best" choice. Obviously, this is simplistic in the extreme since it ignores uncertainties, thresholds and non-linear reactions to equal steps, as well as the possibility of purely qualitative values associated with each choice. Without going into the details, let us briefly sketch a method which attempts only to extract information about preferences which is certain on the basis of the information above. In the simplest version of this, the "pay-offs" believed to be associated with each choice are compared by pairs for each criterion, and a preference matrix is constructed. In the most basic version, this can be done using a simple Boolean response to each question, is choice i better than choice j for this criterion? This gives us the three matrices below for our particular example.

|   | 1 | 2 | 3 | 4 |
|---|---|---|---|---|
| 1 |   | O | O | O |
| 2 | 1 |   | O | O |
| 3 | 1 | 1 |   | O |
| 4 | 1 | 1 | 1 |   |

|   | 1 | 2 | 3 | 4 |
|---|---|---|---|---|
| 1 |   | 1 | O | O |
| 2 | O |   | O | O |
| 3 | 1 | 1 |   | 1 |
| 4 | 1 | 1 | O |   |

|   | 1 | 2 | 3 | 4 |
|---|---|---|---|---|
| 1 |   | 1 | 1 | 1 |
| 2 | O |   | 1 | 1 |
| 3 | O | O |   | O |
| 4 | O | O | 1 |   |

first criterion      second criterion      third criterion

and these can now be combined, by multiplying each by its weighting factor.

|   | 1 | 2 | 3 | 4 |
|---|---|---|---|---|
| 1 |  | $\alpha_2+\alpha_3$ | $\alpha_3$ | $\alpha_3$ |
| 2 | $\alpha_1$ |  | $\alpha_3$ | $\alpha_3$ |
| 3 | $\alpha_1+\alpha_2$ | $\alpha_1+\alpha_2$ |  | $\alpha_2$ |
| 4 | $\alpha_1+\alpha_2$ | $\alpha_1+\alpha_2$ | $\alpha_1+\alpha_3$ |  |

Weight accorded to the three criteria is $\alpha_1$, $\alpha_2$, $\alpha_3$, respectively.

From this final preference matrix a graph may be constructed indicating the relation between the four choices. For example, an individual who considers that the three criteria are of equal importance, $\alpha_1=\alpha_2=\alpha_3$ gives us,

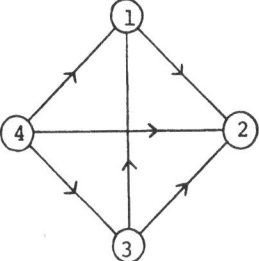

An arrow indicates net preference between the pair.

and clearly, choice 4 is the best. For somebody else, however, if for him the third criterion was very important, more important than the other two combined $(\alpha_3>\alpha_1+\alpha_2)$ then we have the graph,

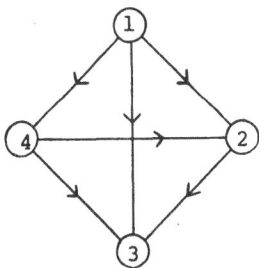

and for this the choice 1 appears as the best.

Of course, this method may be considerably refined to consider weak preferences and a range of indifference, and it can be used to analyze choices where considerable uncertainties exist as to the possible "pay-offs" of each choice along the various axes, as well as for purely qualitative criteria. It has been given the name of ELECTRE[2], and what is particularly interesting for us is that it has been devised in order to attempt to "model" the manner in which individuals really do make decisions, and to capture the uncertainty, subjectivity, ambiguity, and the role of qualitative criteria in people's behaviour.

A particularly clear way of visualizing the problem of choice is to imagine that each actor is at the "origin" of a set of axes each of which represents a criterion involved in the evaluation of the decision. The origin represents the "ideal" choice for the actor with the maximum imaginable pay-off in all directions, and of course has nothing necessarily to do with what that actor actually does, since the real choices presented to him will be somewhere out in the space defined by the dimensions of his value system, a "mental map" of imperfect offer he draws with the information he has received.

In such a space, the "distances" of each choice along any particular axis will be "stretched" or "squeezed" to a degree which depends on the weighting the actor accords to that criterion. In such a representation then, the four possible choices of our previous example viewed by an actor who puts equal weight on each criterion will look like the figure shown below.

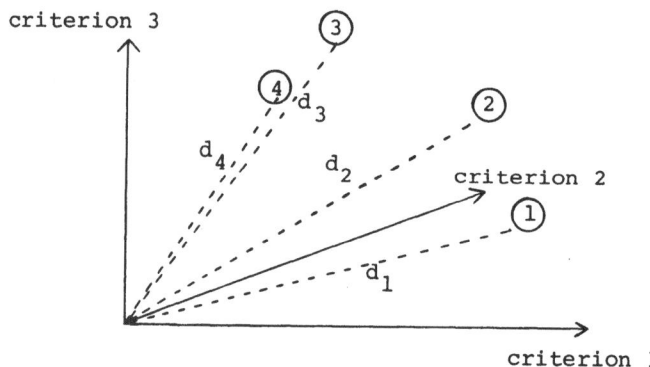

Figure 5. Value space of an actor who assigns equal weights to the three axes. Choice ④ is the most probable because $d_4$ is shortest.

However, the same four choices viewed by an actor who puts a much greater weight on the third factor appears in Figure 6.

Thus, we may view the problem of choice under multiple criteria as the "distance" from the origin, in an n-dimensional space, of the various possible choices. Of course, the position of each point is uncertain to a degree depending on the uncertainties involved in the estimation of the "pay-offs" associated with each choice, and also depending on the degree of precision one can give to the weightings assigned to each axis. Each possible choice is therefore associated

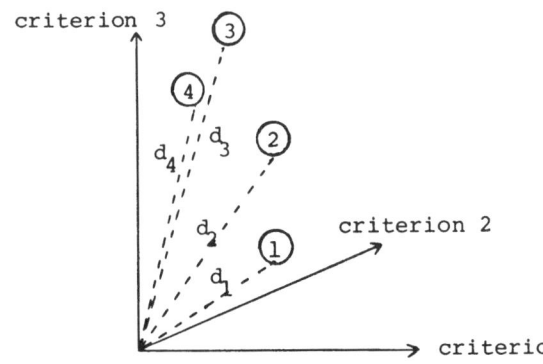

Now $d_1$ is the shortest and hence choice ① is the most probable.

Figure 6. Value space of individual who assigns weights $\alpha_3 > \alpha_1 + \alpha_2$.

with a "cloud" rather than a point, and the problem of decision reduces to that of estimating which choice gives rise to a "cloud" which is nearest to the origin. This corresponds to supposing that the "attractivity" of a given choice decreases as its distance from the origin increases.

Thus, for a single actor at a particular moment we may suppose that the choice i will be selected with probability,

$$\frac{(1/d_i)^I}{\sum\limits_{i'} (1/d_{i'})^I} = \frac{A_i}{\sum\limits_{i'} A_{i'}} \tag{1}$$

where I gives a measure of the informational precision of the distances. Thus, when $I \longrightarrow \alpha$ then the probability of choice is simply 1 for that nearest to the origin, and 0 for the others. In the opposite extreme, of extreme uncertainty, $I \longrightarrow 0$ and we simply have equal probabilities for all choices. Clearly, most decisions fall somewhere inbetween these two extremes.

An important point which we have not yet considered is that of "time". In an evolving system, the "pay-offs" that characterize each choice will change in time, and this evolution will be predicted by the decision maker according to the "system model" he is using. It is somewhat disquieting to realize that the model we are going to build will contain the behaviour of actors, which will depend in turn on the models available to them. However, that's the way it is. He will estimate the distance of each choice at different future times and consider which of these choices, according to the scale of value he assigns to time, is his preferred choice.

If we look now at the behaviour of populations then, assuming that we may define the probability of each individual making a particular choice in an interval is given by the expression (1), then we can construct kinetic equations for the behaviour of the system. If all the decisions made in the system are independent of one another, then we have an essentially trivial problem but if, as is the case in any human system, the decisions that have already been made modify the "pay-offs" perceived by the actors, then we have a far more interesting situation. What occurs is that the choices that have been made, changes in the "real" system, are reflected in the "mental maps" or "internal psychological value spaces" of the actors causing them to modify their behaviour. If we consider, for example, a very simple problem of a homogeneous population which is growing in size, locating around a centre of employment, then initially location occurs close to the centre but later, as the density increases, the choice shifts to more distant locations. In the value space of the population, the choice of the central location while remaining attractive in the dimension of spatial convenience, receded from the origin in dimensions associated with crowding, and led to the adoption of the other more distant locations which seemed more attractive in comparison. Of course, with a single population, and such a simple problem, the evolution is trivial, but if we think of the interplay of decisions that are made by the many different types of actor present in the city, say, we see that we have a highly non-trivial relation between the decisions that have been made, and those that are going to be made, because the dynamic interplay of the system evolution with the value space of the actors is very dependent on the precise timing of events.

Returning now to our urban model, we show in Figure 7 the basic interaction scheme for the six types of population which we have supposed are most important in the evolution of the city. From this, using the assumption that the probability per unit time of an actor making a particular choice is proportional to,

$$\frac{(1/d_i)^I}{\sum\limits_{i'} (1/d_{i'})^I} = \frac{A_i}{\sum\limits_{i'} A_{i'}}$$

then we can construct our kinetic equations expressing the evolution of each variable, in each locality. These are given in Appendix 1, and have the general form,

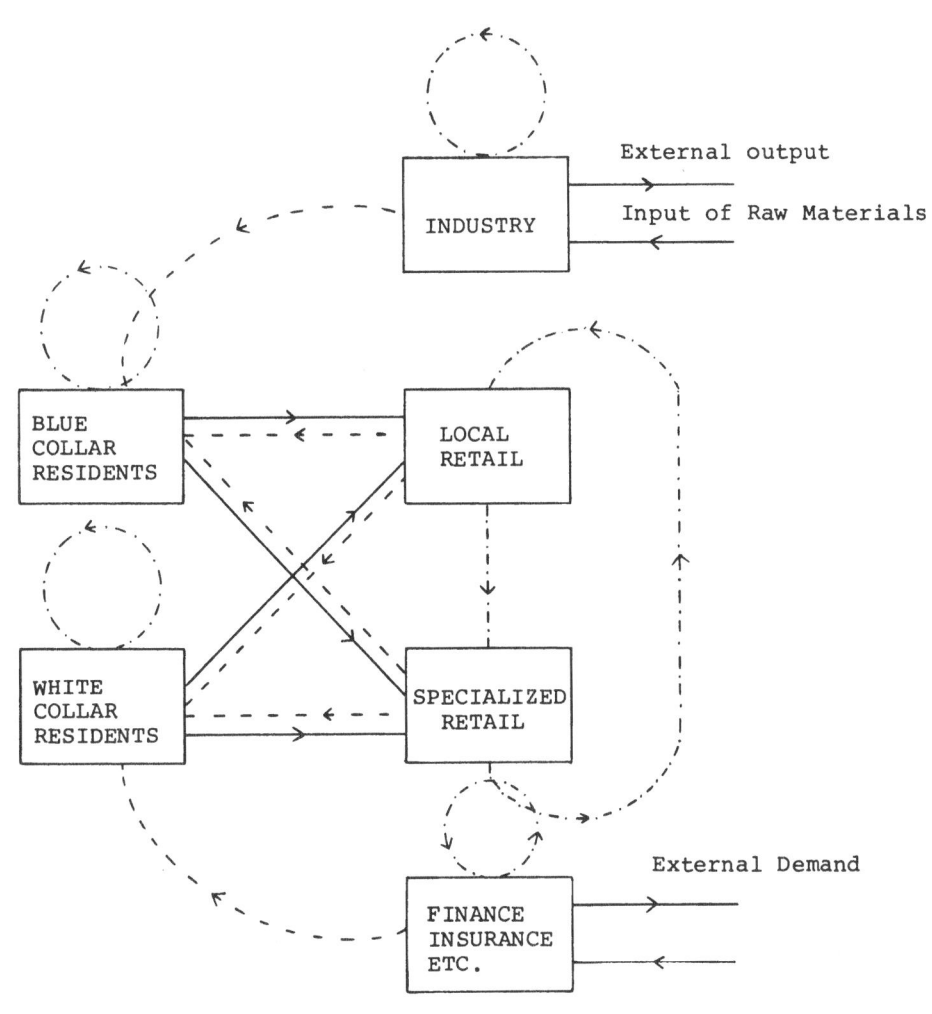

External output

Input of Raw Materials

External Demand

—————————— Demand for Goods and Services

— — — — — — Demand for Labour

—·—·—·—·—·—. Cooperative effects (economies of scale, common
infrastructure, learning, etc.)

Figure 7. The interaction scheme of our simple City System.

$$\frac{dx_i^k}{dt} = (a + bx_i) \left( \Sigma\Sigma J_{jm}^{mk} \frac{A_{ij}^k}{\sum_i A_{i'j}^k} - x_i \right)$$

which expresses how the number of residents of socio-economic group k, at the point i, $x_i^k$, change in time by the residential decisions of the sum of all those k employed in the different possible sectors m, whose jobs are located at j. Thus, $A_{ij}^k$ is the attractivity of residence at i as viewed by someone of socio-economic group k, employed in sector m at the point j. The equation describes the growth of $x_i$ up to the level at which it the fraction of people attracted there is exactly equal to its fractional or relative attractivity. Of course, this latter, $A_{ij}^k$ as we have discussed, is given by some inverse "distance" in the value space of the $J_j^{mk}$. The simple functional form we have supposed for this attractivity, considers basically three dimensions of values, for different possible locational choices. Firstly, we have the effect of distance from the place of employment, which can be fur- ther broken down to allow the consideration of different factors such as time of travel, cost, comfort, etc. In the simulation I shall briefly show, however, we have simply supposed an expression of the type,

$$e^{-\beta d_{ij}}$$

The second factor we have allowed for is the effect of crowding, which of course may be analyzed in terms of price, of the type of building, of noise, etc., and whose effect will vary depending on the particular mix of residents and commercial activities present, since some uses require greater areas than others, and some actors are less sensitive than others to high prices. The functional form used here is:

$$\frac{v^k}{v^k + \sum_m \gamma^m J_i^m + \sum_k \gamma^k x^k}$$

Finally, we have also allowed for the attractivity of a particular point to depend on its natural beauty, and also on the character of the residential population already present. This would allow for the possibility that, for example, people from the upper socio-economic group prefer to live in an area where their own group is already present (it is the mathematical expression of a "nice area").

Putting all these factors together, we find the expression,

$$A_{ij} = \frac{v^k (1 + \sigma^k x^k) e^{-b^x d_{ij}}}{v^k + \sum_m \gamma^m J_i^m + \sum_k \gamma^k x_i^k}$$

which we have supposed to express the value structure of the different types of resident, k, and also how this system of values "reacts" to changing possibilities.

We have written down similar equations for the other actors which in brief express, for example, the need for industrial employment to be located at a point with good access to the outside, and for a very large area per job, as well as some 85% of their workforce being taken to be in the lower socio-economic group. We have also added the fact that the interdependence of many industrial activities leads to a preference for locations adjacent to established industrial locations. This term also covers many subtle effects of the infrastructure that grows around existing situations. The main effects are all noted on the interaction scheme of Figure 7, and the full equations are given in the Appendix, and so here we shall simply proceed to discuss the evolution of our simple city system.

## Urban Evolution

In this section we shall briefly describe some of the simulations that we have made using our simple model. In the first case, we have looked at the evolution of a centre which initially is only a small town but throughout the simulation, due to population growth and expanding external demand from the industrial and financial sectors, the town grows, spreading and sprawling in space as it does, and also developing an internal structure.

The initial condition of the simulation is shown in Figure 8, and the particular values of the parameters which we have used are given in Appendix 2. After 10 units of time, the situation has evolved to that shown in Figure 9, where an internal structure has already appeared. Industry, commercial and financial employment are all still located at the centre, but now we observe residential decentralization, particularly on the part of the upper socio-economic group. The centre is very densely occupied and is strongly "blue collar".

As the simulation proceeds, however, at around 15 units of time, this urban structure becomes unstable. It is not a question of simply growing or shrinking, what is at issue is the qualitative nature of the structure. For, at this point in time, the very dense occupation

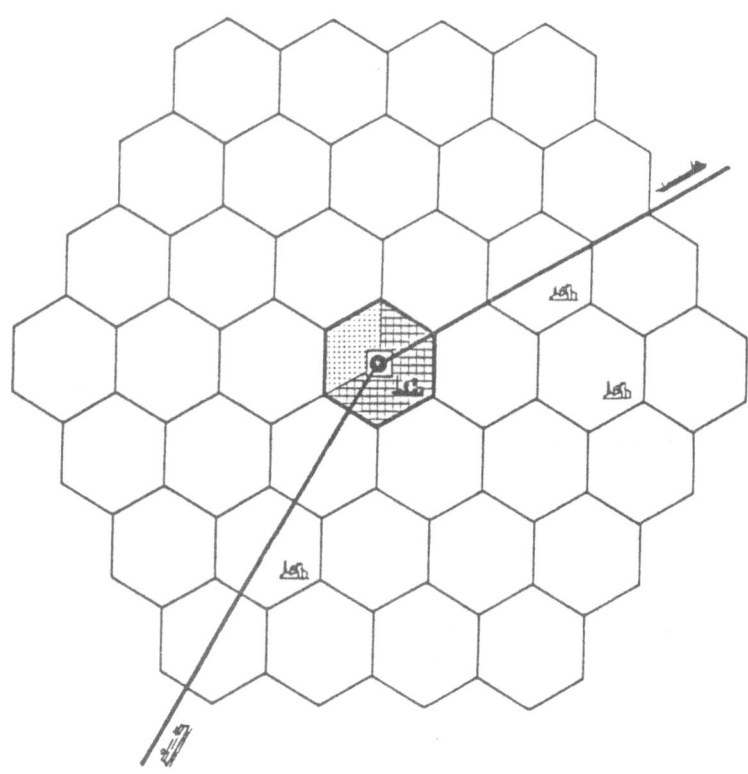

Key: The tightness of the mesh of the square lattice gives the density of "blue collar" residents.

The density of points gives the density of "white collar" residents.

The point size gives number employed in local retail.

The box size gives number employed in specialized retail.

The symbol size gives number employed in industry.

In each locality, represented by a hexagon, the relative numbers of "blue" to "white" collar residents is indicated by the angle subtended by each. Heavy hexagon defines CBD.

Figure 8. The initial condition of our simulation. A small town, unstructured as yet, and lying on a line of communication, begins to grow. The key to the symbols of this figure and those which follow is given above.

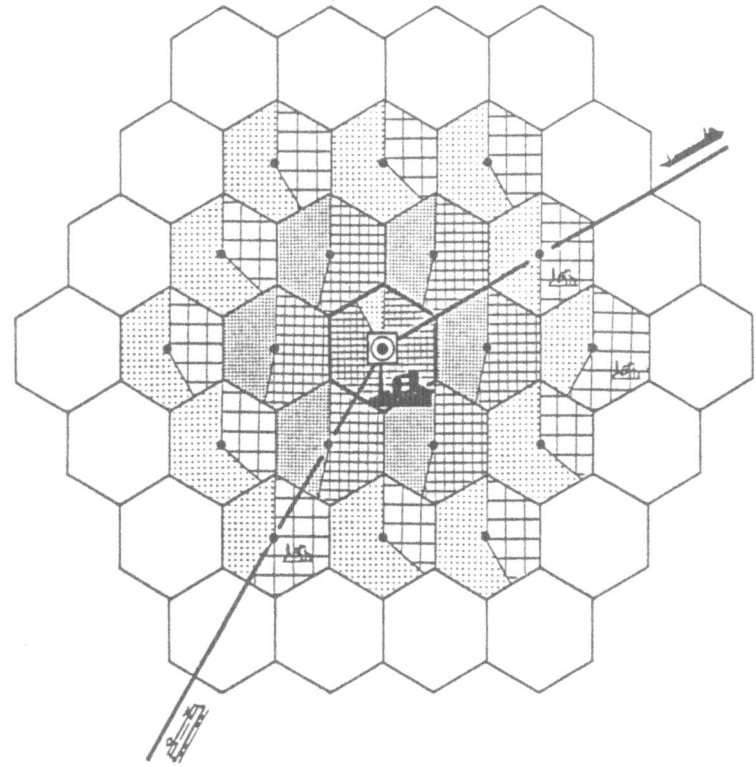

Figure 9. The urban structure after 10 units of time. Residential
        decentralization is already well developed, particularly
        by the white collar workers who have an exponentially
        decreasing density distribution with a crater at the
        centre.  The blue collar workers have a shorter range
        exponentially decreasing distribution, reflecting their
        lower mobility.  The structure is still centralized with
        all employment, except some very local retail, located
        in the centre.

of the centre is beginning to make industrial managers think about some
new behaviour. For some of them the cost of continuing to operate in
the centre is making them contemplate the abandonment of the infrastructure and mutual dependencies that have grown up with time. At
this point, as for a dissipative structure, it is the fluctuations
which are going to be vital in deciding how the structure will evolve.
At some point there is an initiative, when some brave individual decides
to take his chance and to try to relocate at some point in the periphery.
Exactly where will depend on his particular perceived needs and opportunities. However, what is important is that whereas before this time
such an initiative would have been "punished" by being less competitive,
now, around t = 15, the opposite is true. Once the nucleus is started,
and of course its own infrastructure begins to be installed, so almost
all the industrial activities decentralize, and establish themselves
in this new position in the periphery.

In order to show the effects of chance, we had the "seeds"
of an industrial initiative present on several points. By passing
through the instability several times with different simulations, it
was found that minute differences in the relative sizes of the "seeds"
led to the re-organization of industry at different points. However,
owing to the attractivity of the points lying along the communications
axis, it was much more difficult to provoke growth of an industrial
centre away from this axis. Summarizing the effect then, at around
t = 15, the hitherto circular symmetry of the urban system becomes unstable. At this point, many different initiatives could succeed in
carrying the system off to some particular new state of organization.
However, those which succeed with the least effort are the industrial
nuclei in the periphery, lying along the communication axis.

In Figure 10 at t = 20, we see the new structure.

From this point on, however, the locational decisions of the
"blue collar" workers are particularly affected by the fact that their
value systems are now based on the fact that industrial employment has
re-located in the south-western corner of the city. Thus, the spatial
distribution of blue collar residents in the city starts to change,
having in a sense a new focus. This inturn acts on the locational
choices of the white collar workers, who find space easily in the
regions of the city less favoured by the blue collars, and whose spatial distribution adjusts accordingly. Changes in the distribution of
local service employment also occur then, and the whole structure
evolves to the pattern shown in Figure 11, by time t = 40. Here, we
see that we have actually displaced the centre of gravity of the urban
centre, and have an urban structure which resembles two overlapping

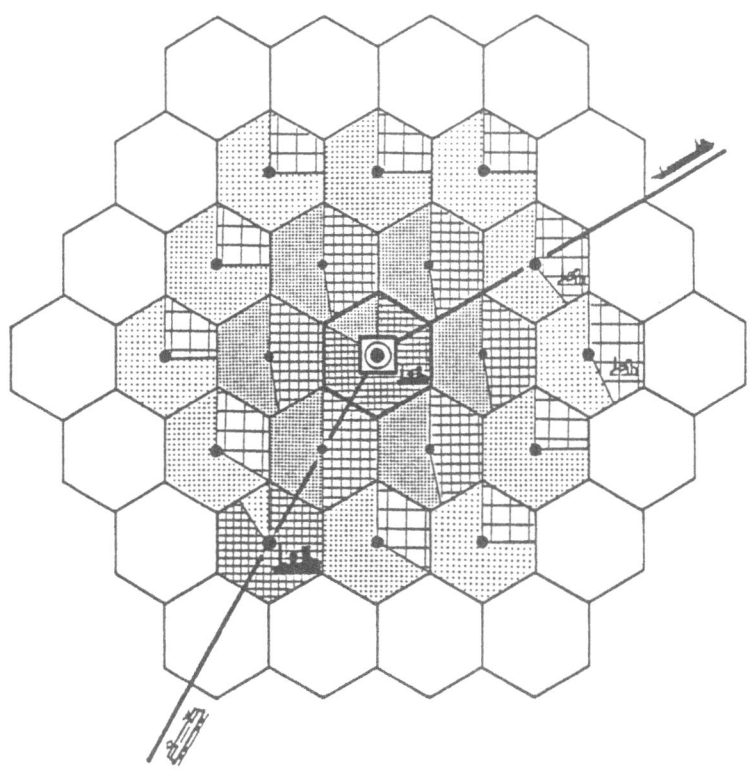

Figure 10. At t = 20 we see that a major re-organization is occurring.
Industrial employment is leaving the centre and re-locating
on the communications axis in the south-west periphery.
Already, the distribution of "blue collar" workers has
changes, affecting the evolution of all the other
variables.

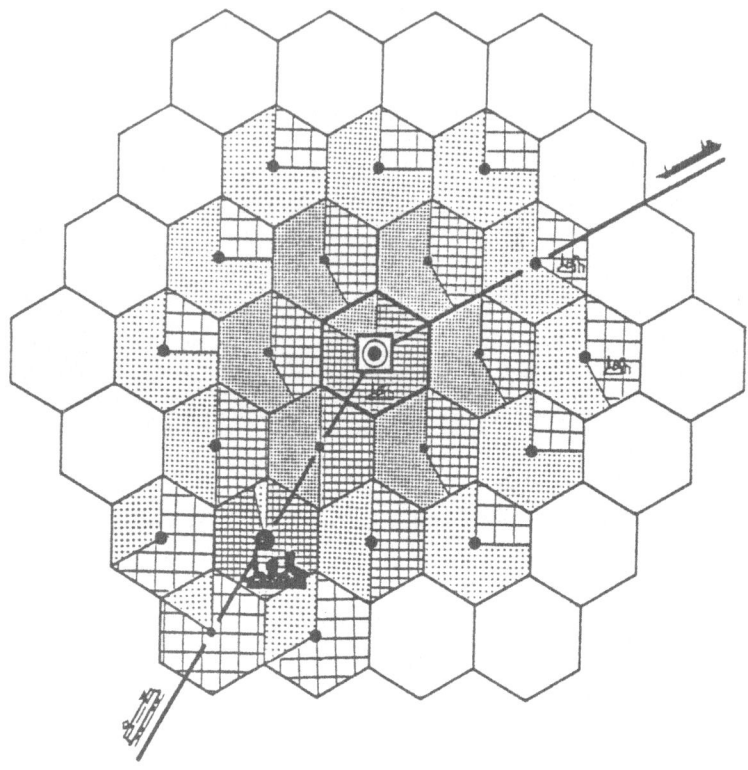

Figure 11. By time t = 40, the urban structure has changed qualitatively from that of Figure 9.  It has developed a second
focus, and has structured functionally.  That is, one
centre is essentially an industrial satellite, while the
traditional centre has become largely a CBD and the important shopping centre.  We may also note that in the traditional centre the retail employment has moved outwards to
the second ring (suburban shopping centres), while in the
industrial centre the retail employment is still
centralized.

urban centres of different character. In the south west we have a
predominantly working class, industrial satellite, while the original
city centre is a CBD and important shopping and commercial district,
with predominantly white collar suburbs stretching away from it on
three sides. In this part of the city, it is the second ring that has
attracted the local shopping centres, while in the industrial satellite
it is the heavily populated, industrial district itself that has
become an important shopping centre.

During the simulation we can calculate a great deal of interesting information concerning the urban structure and its "running
costs". For example, we can calculate the total number of jobs available in each sector and the total travel generated by commuting workers.
This can be calculated separately for "blue" and "white" collar workers,
and if necessary can be calculated for each residential location.
Similarly, from the location and size of shopping centres, together
with the distribution of residences, we can calculate the total travel
involved in consumer shopping trips. Clearly, the energy consumption
of the urban centre is related to the sum of the total travel of com-
muters and shoppers, and this can therefore be calculated. What is
particularly interesting here is that the usual procedure is to simply
divide the total distance travelled in the city by the population and
discuss the average distance travelled per inhabitant. When we look,
for example at the average distance commuting to work of "blue" collar
workers, we find that the trend changes when the city re-structures at
around t = 15, as it also does for "white" collar workers.

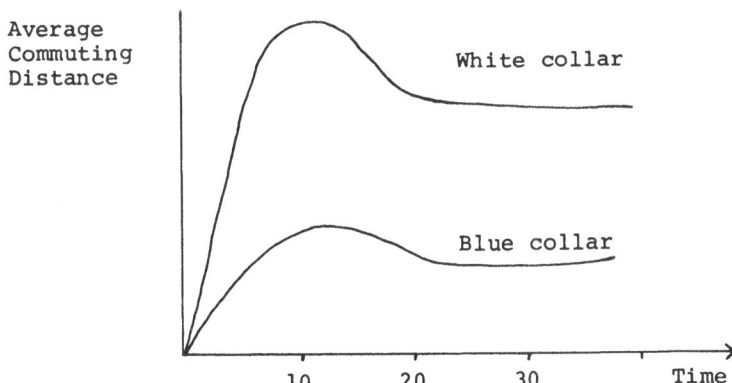

Figure 12. Here we see the changing trend of average commuting
distances due to the structural re-organization of the
system.

This shows us the dangers involved in global modelling, for on that scale what we see is an apparently inexplicable change in behaviour in which the distance travelled per person, and the average energy consumption per person, stops rising and even decreases. Only a model which can describe the internal restructuration of the city could have predicted such a change, and linear systems theory, and input-output flow models would have to be re-calibrated at this point. This also highlights another aspect of modelling method which is sometimes used incorrectly. The important point about, say, the energy consumption of urban travel is that it results from all the travel that is taking place in the city, and hence is an "observable" which has the value it does because the city has the distribution of residences, jobs and shops that it does. It would be quite incorrect to use this total urban travel in order to model the system, or as part of a global model, because as we have seen changes in, say, total urban population can lead, through the type of internal changes that we have discussed, to modified values of the average travel requirements. In other words, relationships between global variables of complex systems are nearly always non-linear and a systems analysis which assumes linearity will only be reasonable in the short term or in a neighbourhood of the calibration.

As a final example here, let us briefly describe a series of simulations which were performed in order to investigate the impact of rising travel costs in a city. In this case, we started from the same initial condition as for the previous simulation but with a slight change in the value of parameter $e^e$ (a systematic examination of the effects of the various parameters is given in Reference 3). At t = 20 the situation is that shown in Figure 13; characterized by circular symmetry, with a CBD, industry and main commercial and shopping concentration in the centre and with "white collar" residential suburbs surrounding it. At this time, we performed two simulations starting from this particular state. In one case we allowed transport costs per unit distance (costs being in time or money) to fall, and the other to rise. The first case corresponds to a policy of heavy investment in order to continue decreasing these costs in a still growing city, while, in the other case, there was perhaps little investment and travel costs were allowed to rise. Possibly also, the first case corresponds to a heavy subsidy on rising fuel costs, and the second to simply passing this on to the consumer.

After running the simulations for a further 20 units of time, the structures which evolved were examined. They are shown in Figure

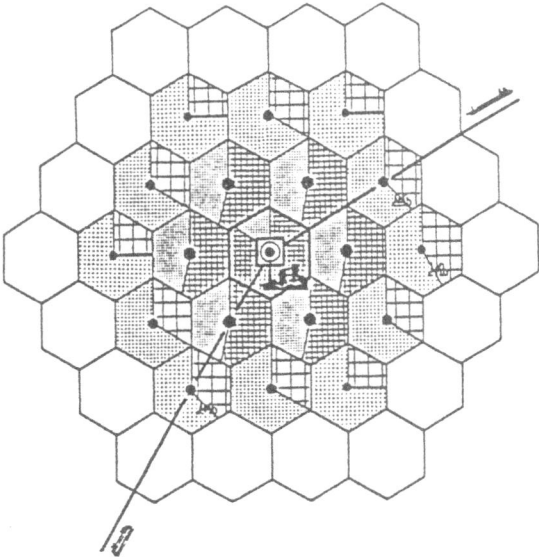

Figure 13.

Low Transport
Costs

High Transport
costs

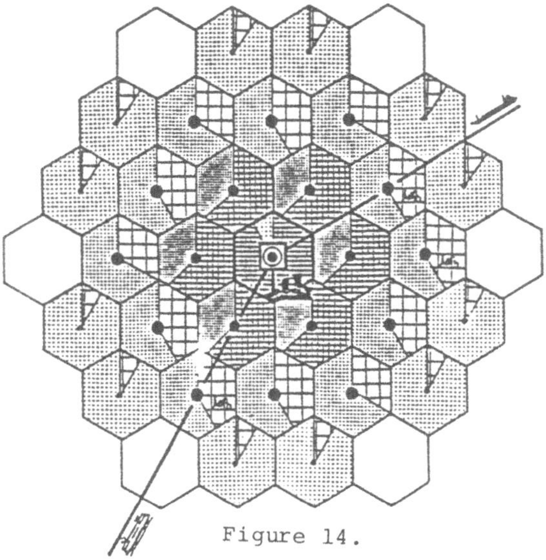

Figure 14.

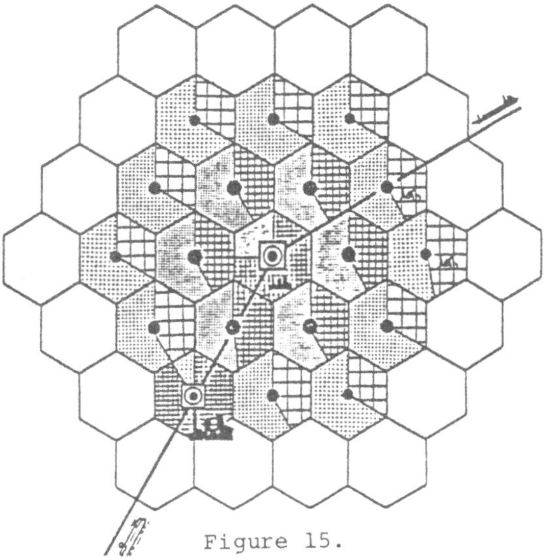

Figure 15.

Two possible evolutionary paths corresponding to decreasing or increasing transportation costs.  The important point is that the decision
results in structural changes which modify the system qualitatively.

14 and Figure 15. We see that they differ qualitatively in that the
simulation performed under falling transport costs still retains its
circular symmetry, while that performed under the scenario of rising
transport costs has become unsymmetrical as industrial activity has decentralized and nucleated in the periphery. The transportation and
energy requirements of the two urban structures are quite different.
The total travel generated in the "low-cost" city is approximately
twice that of the "high-cost" city, and the average commuting distances
of "blue collar" workers is three times as large, while that of "white
collar" workers is doubled. In fact the average "cost" of commuting
for "blue collar" workers is greater in the first case than in the
second. In other words, the reduction in travel cost per unit distance
of the scenario causes a quite different urban structure to evolve,
and this is such that blue collar workers on average must travel much
further to work than in the other case. This means that although the
cost per unit distance decreases, it is more than compensated by the
increase in travel distance that the urban structure requires.

It is interesting to note that the GNP of the city requiring
or generating greater total travel would probably be higher than that
of the second city, although the actual consumption of goods and services is smaller. Again this points out the dangers of using such global
indicators for complex systems, where structure and function are inex-
tricably mixed, and where evolution and changing conditions can lead to
internal re-organizations.

Various other problems can be examined such as, for example,
the effects of regulation of industrial and commercial location, or of
changing patterns of external access for goods and raw materials, changing productivity in industrial or office employment. Similarly, a
study can be made of the effect on the urban structure of the introduction of a metro line, including the chain of events that it sets up in
the long term involving modified land prices, and changing commercial
and residential attractions. Our final simulation shows one of our
preliminary simulations of this problem where we see that apart from
distorting the urban space by causing greater residential densities
along its path we can discern the beginning of two new commercial centres which are forming at each end, and by becoming employment centres
themselves they lead to further modifications of the residential location pattern. Thus the simple decision concerning the building of a
metro line sets off a whole series of events, leading to the formation
of sub-centres, and a change to a polynuclear structure, although in-
tuitively the effect of a metro line running to the centre of town
would be to reinforce and preserve the status of the latter.

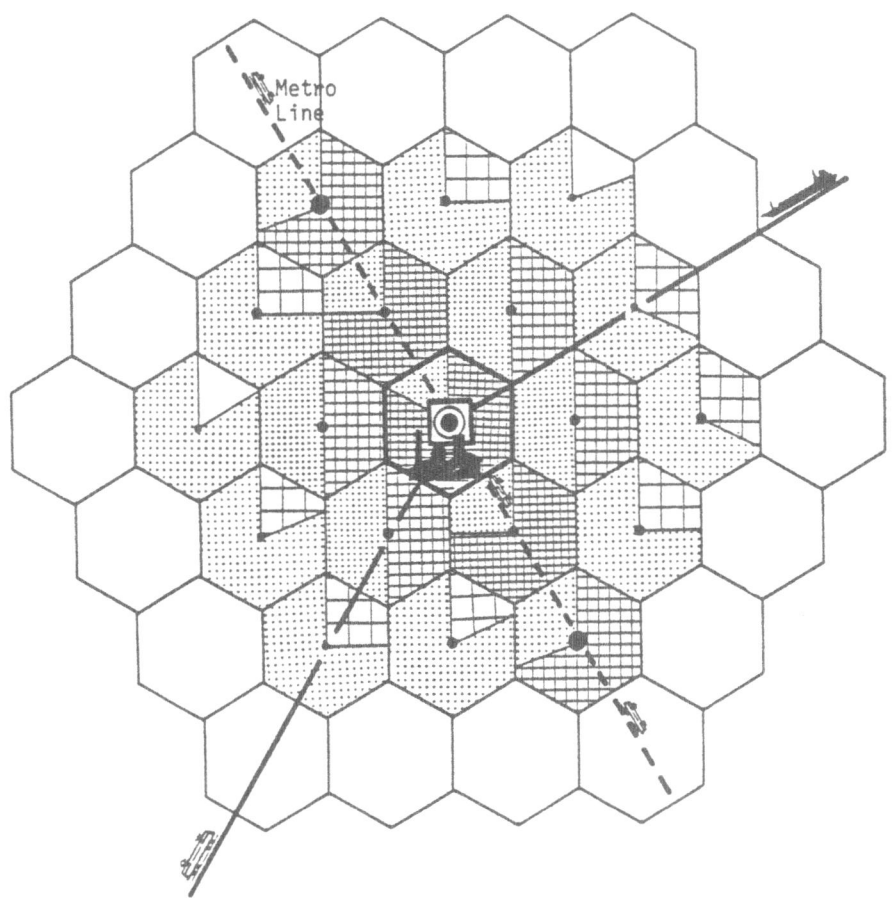

Figure 16. The evolution of the urban system is modified by the
decision to build a metro line. Apart from concentrating
the population densities along its length, causing changes
in land prices, etc., as it does, and changing the residential mix in consequence, it also provokes the growth
of two sub-centres of retail employment at its outer ex-
tremities. This could be very important in later
evolution.

## Conclusions

The simulations which we have briefly described illustrate the self-organization of our simple urban system. They show how policy decisions concerning transportation, housing regulations, industrial location, etc., can modify the evolution of the system qualitatively, leading to new spatial patterns and to changes in the trends of macro-variables related to the whole city.

Of course, a city model can only be used to explore future possible evolutions by placing it in its regional and national context, and a similar model, based on the analogy with "dissipative structures", has been developed to describe the evolution of a region and of an urban hierarchy as a result of such, historically dependent, self-organization processes[4].

An important general point that arises from such models is that a structural re-organization of the urban space leads to a corresponding re-organization of the mental maps and values of the various actors. Essentially the symmetry breaking properties of self-organization lead to a corresponding expansion of the dimension of the actors' value space. For instance, in the example given above, while initially under circular symmetry, the variables and parameters of decisional criteria can all be expressed in terms of the scalar distance from the centre, once the symmetry is broken, the value space expands to include all the angle dependent possibilities. Similarly, for example, when all cars were black the question or value attached to colour was of no importance. Once the symmetry had been broken, however, and cars of other colours appeared, then the new dimension is created in the value space of buyers, and finally it can become an important factor in sales.

Complexification feeds on itself because it creates new situations and dimensions which widen the experience of people and create new tastes and qualities, leading to new behaviours and to further complexification.

The important point is that fluctuations around "normal" behaviour in the real world and fluctuations in the mental models of actors both explore situations which are "richer" than the reduced description of the world which is given by our model. These explorations can be amplified by the non-linearities in the system and lead to a structural evolution of the system. But is all this really true or is it wild speculation? Where precisely are all these non-linearities which I am suggesting characterize human systems so strongly?

Firstly, there exist purely physical non-linearities in the

workings of objects, related ultimately to energy and matter flows, but also to such effects as, for example, surface area to volume ratio. These lead to "optimal sizes" for elements, and give rise to economies and diseconomies of scale, to division of labour, to aggregation and cooperation, pooling of resources, etc. Let us try to imagine a "city" of say a million inhabitants which has no internal structure. Each small locality contains small units of each type of industry, of all types of shops and services. In such a city there are no "head offices" or central depots because such things already arise because of the spatial self-organization processes of the type I am attempting to imagine absent. Amid all this, we find at each point the same mix of blue collar and white collar residences. Clearly, such a vision is impossible in reality because real advantages exist in the functioning of certain size units of each type (due to analogous questions of internal organization), which means that they obey non-linear laws. Thus, any small fluctuations away from total uniformity will be amplified by the advantages perceived by at least some of the actors. Herein lies the other very strong source of non-linearity in human systems. It is the term:

$$\frac{(1/d_i)^I}{\sum_{i'} (1/d_{i'})^I}$$

which we derived from our model of preferences which is inherently and dramatically non-linear. If information is readily available, I is large and even a small change in the advantage offered by a particular choice i can result perhaps in a large change of the population's behaviour. Thus, it is the human capacity to "treat" information and to choose his behaviour which is at the root of much strong non-linearity. However, even in a system with rather poor information concerning the "true" advantages and disadvantages of different behaviour, which I believe is our situation, people's behaviour is largely determined by repeating previous actions which were not calamitous, and when a change is thought desirable, by imitating others. This imitation introduces a strong element of non-linearity into the problem.

Equally true, of course, is the fact that by manipulating information one can change the evolution of the system. Both direct advertising and propaganda as well as social pressure in the form of fads and fashions can create desires and frustrations which may mark the system permanently. Values, it seems, are not the simple, self-evident certainties which we may have believed. Even such "sure-fire" values as maternal love have recently been revealed to be time dependent /5/. What we must face is that almost all of our everyday actions are not the expression of an absolute rationality, but the result of a dynamic dialogue between "system" and "values", between "supply" and "demand", during which bifurcations occur. Their "rationality" is simply conferred on them by the society in which they are thought normal, where they evolved, and they can, and will, change. The problem of policy making in a world with changing values is indeed a fundamental one.

  Summarizing the main points made above, then, I believe that the true nature of living systems is only beginning to be understood. The new concepts arising from "dissipative structures" offer us a new basis for understanding the structure and function of systems. Thus, when we approach a complex system and wish to model it, if our "system structure" simply reflects the structure and function present at the initial moment and we then calibrate the "model" on the initial state, then any prediction that the "model" makes must automatically assume that the function and structure do not change. This may lead to predictions which are not only quantitatively wrong but also qualitatively wrong. What the new paradigm of dissipative structures tells us is that the initial structure and function of the system was itself the result of an evolutionary process which, after a particular history involving both micro and macro factors, was established in the system. Because of the existence of multiple solutions, in fact, the dynamic equations of the macroscopic variables are ambiguous and could have given rise to "other" structures if the particular history had been different. That is a different series of fluctuations around the average values. Thus, if we admit that the particular initial structure and function with which we start is a "special case" and that micro factors "outside the model" led to its establishment, then we must also admit that this will be true of the future evolution. Any "model" therefore which takes the initial structure and function as its basis is bound to fail when the first of these fluctuations carries the system to some new state of organization. In order to build models which can cope with such problems, we must therefore look for the underlying interaction processes which can give rise to the many different structures and function that we may observe in different circumstances. Thus, for our chemical system it is the same unchanging reaction scheme of the Brusselator which gives rise to either stationary or moving coloured bands, spiral moving waves of different types or clock-like oscillatory behaviour under different circumstances and under the same circumstances but different histories. Similarly, the same basic interaction scheme of the figure can give rise to cities with extremely different structures and traffic flows, energy consumptions and neighbourhood qualities.  It is not proposed as a serious model of a city, but rather as an illustration of the sort of model that should be developed, and the sort of study that should be made in order to explore the qualitative results of different decisions and events.

Of course, in the real world of pressing problems where decisions must be made now, then it is probably better to perform a global systems analysis than not to, and therefore such models should not be discouraged.  However, parallel with such activity, there must be a great research effort aimed at developing the sort of deeper understanding of function and structure in non-linear dynamic systems which I have discussed here.  Our models should not simply say: the system is organized like this.  They must also examine the question of why is it organized like this?  The reply will involve an understanding of the stability of this particular structure, and in turn this will allow an appreciation of its potential instabilities, and of the new dimensions and levels of organization that may be created in its future evolution.

## Acknowledgements

The author wishes to thank Professor I. Prigogine whose ideas have inspired this approach for his constant interest and invaluable comments.  The urban models were developed in collaboration with Mms. F. Boon and M. Fischer-Sanglier, and owes much to the discussions with and support from R. Crosby, D. Kahn and F. Hassler at the Department of transportation USA.  This work was also supported by the Actions de Recherches Concertées of the Belgian Government, under convention no. 76/81 phase II.3.  The author is also most grateful to P. Kinet for drawing the urban simulation figures.

## Appendix 1

The different types of equations expressing the evolution of these variables in each point of system are:

a)
$$\frac{dS_J^E}{dt} = \varepsilon^E S_J^E \left( D^E \frac{A_J}{\Sigma_J A_J} - S_J^E \right) \qquad (II.1)$$

with

$$A_J = \left[ \frac{(1 + e^E s_J^E)}{(\mu^e + \alpha_J \phi^E)} \quad x \quad \frac{\zeta^E}{(\zeta^E + \sum_{k'} x_{J'}^{k'} + \sum_{e'} s_{J'}^{e'})} \right]^{co^E}$$

this equation describes the evolution of the employment linked to an external demand which we call the "industry" $(S^1)$ and the "finance" sectors $(S^2)$.

b)
$$\frac{ds_J^u}{dt} = \varepsilon^u \ s_J^u \ (\sum_{J'} \frac{\beta^u \sum_{k'} x_{J'}^{k'}}{\mu^u + \phi^u \delta_{JJ'}} \quad x \quad \frac{A_{JJ'}}{\sum_{J*} A_{J*J'}} - s_J^u)$$

with

$$A_{JJ'} = \left[ \frac{(1 + \sum_{e'} e^e s_{J'}^{e'})}{(\mu^u + \phi^u \delta_{JJ'})^{c^u}} \quad x \quad \frac{\zeta^u}{(\zeta^u + \sum_{k'} x_J + \sum_{e'} s_{J'}^{e'})} \right]^{co^u}$$

this equation describes the evolution of the employment related to a local demand which we call the "ubiquitous" $(S^3)$ and "specialized" servi/ces $(S^4)$.

c)
$$\frac{dx_J^k}{dt} = \eta^k x^k \ (\sum_{J'} \sum_{1'} \xi^{kl} s_{J'}^{e'} \frac{R_{JJ'}}{\sum_{J*} R_{J*J'}} - x_s^k)$$

with

$$R_{JJ'} = \left[ \frac{\partial^k (1 + \tau^k x_J^k) e^{-G^k \delta_{JJ'}}}{\partial^k + \sum_{k'} x_{J'}^{k'} + \sum_{1'} s_J^{e'}} \right]$$

this equation describes the evolution of the two types of population considered, which we have called "blue collar" residents $(x^1)$ and "white collar" residents $(x^2)$.

## Meaning of the parameters

The $\varepsilon^L$ and $\eta_.^k$ represents the constant of the rate of reaction of the employments L and the population k to external environment.

$A_J$ and $A_{JJ'}$ represent the attractivity of the function at the point J.

$D^E$ is the external demand for the function E that in the following simulations we have kept constant.

$e^L$ measures the cooperativity between the different functions.

$\mu^L$ production cost which contains the input cost for the industrial sector.

$\phi^L$ transportation cost.

The $\alpha_J$ parameter represents the access of the point J for the communication routes and in these following simulations it is more especially the access to the canal/railway which is very important for heavy industry.

The $\tau^L$ parameter measures the intensity of the crowding supported by the function L, the crowding of a point J take into account the population $\sum_{k'} x^{k'} x_J^{k'}$ and all the employments $\sum_{l'} x^{l'} s_J^{e'}$ at the point J.

$\beta^{u*}$ is the quantity of function u demanded per individual at unit price.

The parameter $CO^L$ measures the unanimity of the response of the consumers.

$e^u$ is the elasticity of the service u.

The parameter $\xi^{kl'}$ is related to the percentage of people of the type k working in the sector $l'$.

$R_{JJ'}$ is the residential attractivity of the point J viewed by someone who is employed at the point J'. It contains the parameters $\tau^k$ which expresses the affinity between members of a population of the same type.

$\partial^k$ represent the sensitivity to the crowding perceived by the population k. The parameter $G^k$ is related to the ease with which an individual of the type k may commute daily the distance $\delta_{JJ'}$ which is the distance between his residence located at J' and his work at point J.

References

1. P. Glansdorff and I. Prigogine, (1971) "Structure, Stability and Fluctuations", Wiley Interscience, London.

   G. Nicolis and I. Prigogine, (1977) "Self-Organization in Non-Equilibrium Systems", Wiley, New York.

   I. Prigogine, P.M. Allen and R. Herman, (1977), "The Evolution of Complexity and the Laws of Nature", in "Goals for a Global Community", Eds. Laszlo and Bierman, Pergamon Press, New York.

2. B. Roy, (1968) "Classement et Choix en présence de points de vue multiples" (La Méthode Electre). Revue Française d´Informatique et de R.O. 8, 57-75.

   B. Roy, P. Vincke and J.P. Brans, "Aide à la Décision Multicritère", (1975). Revue Belge de Statistique, d´Informatique et de R.O. Vol. 15, 4, 23-53.

3. P.M. Allen, M. Sanglier and F. Boon, "A Dynamic Model of Intra-Urban Evolution", Second Interim Report to the Department of Transport USA under contract no. TSC-1640 (1980).

4. P.M. Allen, J.L. Deneubourg, M. Sanglier, F. Boon and A. De Palma, Reports to the Department of Transportation, USA, under contracts TSC-1185, TSC-1460 and TSC-1640.

   P.M. Allen and M. Sanglier, "A Dynamic Model of a Central Place System", Geographical Analysis, Vol. 11, no. 3. 256-272. (1979)

   P.M. Allen and M. Sanglier, "A Dynamic Model of Urban Growth - II", Journ. Social. Biol. Struct. no. 2, 269-278 (1979).

5. E. Badinter, "L´amour en plus", Flammarion, Paris (1980).

# THE IMPACT OF ENERGY AND ENVIRONMENTAL POLICY
## ON THE DESIGN OF ENERGY MODELS

Siegfried K. Gehrecke

Faculty of Organizational Sciences,
11090 Belgrade, Yugoslavia.

## Introduction

In modern industrial societies energy supply for the diverse
activities of the national economy is performed by complex energy systems such as, for example, crude oil extraction and refining, coal
mining, electric generating and distribution.  In the past, energy
systems grew more or less spontaneously like other industries, on the
basis of individual industrial planning and profit-directed calculations.  Government activities in the energy area usually kept in the
frame of general economic policy and most related to the regulation
of particular energy markets, leasing of public lands and off-shore
areas and the sponsorship of research and development on advanced
technologies.  Consequently, energy policy was identical with the
business policies of the major energy industries.

Of late, rising existential problems of mankind, like energy
crisis and environmental crisis, have begun to change this situation.
The awakened public interest for energy and environmental problems
inevitably leads to a more intensive participation of federal, state
and local government in energy and environmental policy formulation
and energy systems planning.  Today, energy policy making is no more
an exclusive concern of the energy industries only, but also a concern
of the public sector and of spontaneous citizen groups.  All government levels are becoming increasingly involved in the processes of
defining the goals of a rational energy policy and of planning, stimulating and coordinating all activities to achieve these goals.
Clearly the goals of energy policy must be embedded in the superior
goal system of the society which includes economic, environmental and
social goals, too.

At all times, energy models have been of great utility to
energy policy formulation and energy systems planning.  By definition
of Charpentier, the meaning of the term "energy model" must be very

broad and not confined to econometric approaches. Each tool which can contribute to a better understanding of the complex energy system area could receive this designation /1/. In this paper, however, the term "energy model" will be confined to quantitative analytical techniques. The Jülich-workshop on Energy Modelling classified the quantitative analytical techniques used for energy systems analysis and policy development into three main groups:

- optimization approaches,
- simulation approaches, and
- econometric approaches /2/.

The econometric approaches have especially been used for a long time in the energy industries for projecting future energy demand and for planning additional capacities. The design of such econometric models is given by the business policy of the interested firms. A very simple and well-known example is the so-called "rule of doubling electric energy demand every ten years" which has for a long time influenced the investments of the electric utility sector in many countries and which still today can be heard in atomic energy discussions.

The intention of this paper is not directed to the analysis of energy models designed by business policies and dependent on the goal-systems of partial groups but of complex models which involve the whole energy system of a national economy observed from a superior point of view, especially of optimization models for the overall energy system embedded in the triad of economy, society and environment. Obviously, the discussion of the impact of energy and environmental policy on the design of such models can for several reasons be neither complete nor exhaustive. At first, there is a large range of available approaches and quantitative techniques applicable to this area as, for example, linear programming and its extensions, dynamic programming and its derivative, static and dynamic input-output analysis. Secondly, there already exists a great number of energy models designed at various levels of problem-insight and model-development and using different techniques and analysis-mix. As important models for complex energy planning, I would especially mention the LP-models for optimizing the energy structure of a national economy established by H. Požar from Zagreb, Yugoslavia /3, 4/ and by K.C. Hoffman from Brookhaven, USA /5, 6/. Thirdly, the involvement of environmental, economic and social aspects in the energy systems analysis opens an unlimited field of new ideas and approaches beginning with the analysis of goal systems and socio-technical scenarios up to the evaluation of social cost and benefits and the quantification of intangibles.

Consequently, the following sections of this paper deal with some specific questions met during the author´s work in the field of energy modelling.

## The Range of Energy Policy in Energy Modelling

The range of energy policy in complex energy system modelling is determined by the relative position of the energy goals within the overall goal system of the society, especially by the relations between energetic, environmental, social and economic goals. Clearly, the significance of energy depends on the degree of its scarcity, respectively on the intensity of the actual energy crisis. In the case of serious energy supply shortages and a threatening collapse of basic economic activities, energy goals would surely have an absolute priority and energy models another design. In the transportation sector, for instance, the optimization procedure could involve not only substitution processes among competitive energy forms for a given fuel demand, but also substitution processes between alternate propulsion systems and between private and public transportation systems for a given transportation demand, by criteria of specific energy consumption or cost minimization. In this case the energy flow network in the Požar-model had to be completed by nodes for the additional substitution processes and extended up to the propulsion energy demand of the diverse transportation systems or to the transportation demand of the main transportation categories such as urban or overland transportation of persons or goods. There is no problem in designing such energy models; the problem is, however, to define the range of energy policy and its relation to urban or national transportation systems policy, employment policy, environmental policy and other important areas of a society´s goal system.

Let us now consider the following policy problem discussed by W. Lieb /7/. Combustion engines are still the most used propulsion systems for private automobiles. The largely dominating Otto motor runs on gasoline, the Diesel engine on diesel fuel. Compared with a gasoline-driven automobile of the same power

- the purchase of a Diesel is more expensive;
- the Diesel engine is working at a higher level of technical efficiency and consequently with lower fuel consumption and operating cost /8/;
- the emissions of a Diesel engine are lower and less harmful;
- the driving characteristics of a Diesel such as accelera-

tion and motor noise are usually somewhat inferior.

Lieb discussed the question as to how far the choice among the above propulsion systems belongs to the area of energy policy and how far to the areas of economic, social and environmental policy. According to his opinion, the driving characteristics of a combustion engine represent an essential attribute of the individual standard of living in the transportation area. Starting from the social constraint of non-decreasing standard of living, Lieb limits the competence of energy policy in this question to substitution processes among equiva-lent combustion engines. Assuming that the new generation of Diesel engines developed by Volkswagen and other companies fulfills these conditions, energy policy can include this substitution process into the optimization procedure for the future overall energy system. In this case, the network of possible energy flows in the Požar-model (Figure 1a) can be widened by an additional substitution node at the right-hand side (Figure 1b). In an analogous way demand equations have to be widened, too. It should be noted that energy demand in the new equation is expressed in terms of propulsion energy (useful energy) and that the equation should involve the technical efficiencies of converting fuel to useful energy.

On the other hand, the shown widening of network and demand equations should not be accomplished if the criterion of equivalent driving qualities is not fulfilled. In this case the design of the Požar-model (Figure 1a) will not be changed and the choice between the automobile propulsion systems will be withdrawn from the competence of energy policy modelling.

Substitution among private and public transportation systems is a more sensitive problem which has not yet had the chance to be treated under the aspect of energy cost and efficiency. Therefore, energy models should not today be extended to the energy-economic op-timization of those alternatives.

## The Relative Importance of Economic and Environmental Goals in LP-Models with One Dimensional Objective Function

It is well known that air pollution and thermal water pollu-tion are for the most part caused by the diverse energy processes, beginning with energy resource extraction up to useful energy consump-tion. Therefore, it is no wonder that energy policy becomes more and more confronted with environmental requirements. Awakened public in-terest for pollution problems has induced a rising significance of environmental goals in the overall goal system and increased involvement of environmental aspects in energy system optimization models. The standard approach to this problem used by K.C. Hoffman /5, 6/, H. Požar /4/, R. Thoss /9/ and others, is:

- to assign emission coefficients for the main pollutants to each energy form and conversion, delivery and utilization device and to evaluate the total emissions for alternative options;
- to introduce environmental objectives formulated as emission constraints for the pollutants under consideration.

The usual objective in the objective function of the standard approach is cost minimization. Considering the formal construction of the standard approach, some authors have tried to draw conclusions about the relative importance of economic and environmental goals represented by the objective function and the environmental constraints. According to Döllekes' opinion, for instance, the involving of environmental constraints in the optimization model means that in the overall goal system of society environmental goals have a higher priority than economic goals /10/. Döllekes argues that the environmental objectives placed in the constraints have absolute priority with respect to the economic objective represented by the objective function, because in the optimization procedure the constraints must be fulfilled absolutely while the objectives in the objective function will be realized only as well as possible. Thoss remarks, that "it is often not appreciated that the most important goals must be formulated as restrictions and that only the least important goal may be used as an objective function" /11/.

This argumentation which is reasonable at a glance, contains an inadmissible interpretation of goal priorities by formal mechanisms of the mathematical procedure. Hax has pointed out that in the LP-problem the objective function and the constraints represent a unity of equivalent conditions for evaluating the optimal solution /12/. With regard to the formal aspects of the LP-procedure no one of these mathematical components could get the attribute of greater importance. Consequently, the mathematical structure of the optimization problem cannot give any criterion for determination of goal priorities.

Indeed, in the standard approach goal priorities are determined in another way, and that by fixing the level of the environmental objectives placed in the constraints. Given cost minimization in the objective function, a low level of admitted emissions increases the relative importance of the environmental objectives and decreases the relative importance of the economic objectives. On the other hand, fixing a high level of admitted emissions decreases the relative importance of the environmental objectives and increases the importance of the economic objectives placed in the objective function. Formally, this procedure can be seen as a simple version of multiobjective optimization in which the optimal compromise among competitive economic and environmental goals must be found in the political process of fixing environmental objectives. The standard approach realizes that compromise at the level of given environmental constraints and evaluates the best suited option (see Figures 2a and 2b).

Compared with more sophisticated multiobjective programming techniques, the standard approach has a very simple design. According to my opinion, this is an advantage rather than a handicap in the analysis of energy and environment for several reasons:

1) Unlike cost, emissions of different pollutants cannot be added and particular objectives must be established for every pollutant, e.g., for $SO_2$, $CO$, $CO_2$, $NO_x$, $C_mH_n$, particulates, waste heat, radioactive waste, and that for air pollution and water pollution separately.

2) Analysis of most environmental problems must be focussed on the regional pollution level which means that for every region particular emission objectives must be established dependent on the specific geographic and climatic characteristics of the region /13/.

3) The inevitable diversification of environmental objectives by the above criteria generates a multitude of objectives that can more easily be formulated as a set of constraints than as an aggregate objective function.

4) In society, environmental objectives are usually calculated in terms of constraints, and that in terms of imission and emission standards. The use of analogue objective constraints in energy modelling increases the transparency and intelligibility of the decision problem for both the energy system and analysis groups and those with policy responsibility and decision competence. The meaning of a higher $SO_2$ level in an urban region can better be understood than the meaning of changing ponders in an aggregate objective function.

5) Finally, the use of the standard approach as an interactive decision model enables all groups concerned in energy and environmental policy problems to evaluate the economic consequences of alternative levels of environmental constraints, to discuss the opportunity cost of environmental policy measures, to revise the former goal priorities and to develop more appropriate environmental standards /14/.

## The Compatibility of Goals and Means in Energy and Environmental Policy Modelling

Describing the objective function of his optimization model, K.C. Hoffman notes that "social costs, or externalities, may also be included if these can be quantified." /5, p. 127/ What is the meaning of this statement when applied to the above mentioned standard approach of energy-economic and environmental optimization?

In economic theory, the concept of "social cost" relates to negative side-effects of production and consumption activities in the national economy. Air pollution and water pollution and their effects on man and his environment are a classical example for it. The main characteristics of social cost are interdependency and lack of compensation: one person's behaviour creates a cost to other persons, but the one who creates the cost is not made to pay for it. Compensation of losses and damages must often be paid by society as a whole.

Consequently, social costs of pollution processes are usually defined as damage costs as, for example, cost of ill health or invalidity caused by air and water pollution, or as compensation costs as, for example, cost of downstream water works established for converting polluted water into drinking water. Before introducing such cost categories into the objective function of the standard approach, the model designer should analyze the following questions:
- What are the economic and environmental goals represented by the model?
- Does the structure of the economic and environmental objectives formulated in the objective function and the objective constraints agree with these goals?
- What kinds of means are available to achieve the goals, and how can they be formulated in the model?
- What are the repercussions of different means upon objectives design?
- Which combination of objectives and means is best suited for the given purpose of the model, with respect to the logical unity of goals and means in the sense of Max Weber?

Regarding the problem of involving social cost into the objective function of standard-approach-designed energy system optimization models, the following answers could be given:
- The economic goal is to maximize welfare, the environmental goal is to maximize environmental quality.
- The objective of cost minimization agrees with the economic goal under the condition of including existent social cost or removing the

existence of social cost. The environmental goal can adequately be
expressed by reduction of emissions to the targets established in
the constraints.

- The requirement of emission reduction to the given environmental
  targets limits the system analyst´s choice among available means to
  those which can ensure diminution or avoidance of pollution processes.
  In the standard approach these means can be formulated as alternative
  processes with improved emission characteristics and usually somewhat
  higher cost. On the other hand, means for compensation or restora-
  tion of damages without changing the pollution level have no relation
  to the given environmental objectives.
- The repercussion of means for avoidance of pollution upon the economic
  objectives is that they remove the existence of social cost by converting external damage cost to internal avoidance cost of the pollu-
  ting processes. In this case, the damage cost version of the social
  cost concept is replaced by the avoidance cost version. The economic
  objective is now to minimize overall cost including the avoidance
  cost of available alternative technologies.
- The use of the "avoidance means concept" to achieve the environmental goals and of the "avoidance cost concept" to achieve the economic
  goals ensures the logical consistence of the optimization model and
  a high level of compatibility of established goals and means.

The interrelationship between goals and means discussed
above is shown in Figure 3. The closed circle of full lines represents
the consistency of the avoidance concept; the alternate concept of the
damage cost version is suggested by interrupted lines.

Considering K.C. Hoffman´s statement about possible involvement of social cost in the objective function, we must draw the conclusion that such involvement in the standard approach should be done
in the indirect form of avoidance cost, and that for reasons of data
availability as well as for reasons of the logical consistence of the
model. In the damage-cost-concept, data availability is a hard problem not yet solved satisfactorily. Cost assessments in this area tend
to underestimate the relative importance of non-damaged man and
environment for the overall quality of life /15/.

## Conclusion

The problems discussed above represent only an arbitrary
selection among questions met in energy systems modelling. Analysis
of these problems could be widened in many directions, e.g., to the
question of consistency of avoidance cost and damage cost in a mixed
social cost concept if such is possible.

As already mentioned above, the intention of this paper was not to give an exhaustive and complete discussion of all possible impacts of energy and environmental policy on the design of energy models, but to stimulate similar reflections in all groups concerned in energy system analysis and modelling.

## References and notes

1. Charpentier, J.P., Overview on Techniques and Models used in the Energy Field. Arbeitsseminar: Energiemodelle für die Bundesrepublik Deutschland, Jülich, 1975.

2. Arbeitskreis 3: Methodische Probleme bei der Erstellung von Energiemodellen. Arbeitsseminar: Energiemodelle für die Bundesrepublik Deutschland, Jülich, 1975.

3. Požar, H., Un modele mathématique pour déterminer la structure optimale de l´energie. Commission économique pour l´Europe (ECE), Colloque sur les modeles mathématiques des secteurs de l´économie energetique, Alma Ata (USSR), 1973.

4. Požar, H., Matematički model za optimizaciju energetske strukture. Institut za elektroprivredu, Zagreb, 1979.

5. Hoffman, K.C., A Unified Framework for Energy Systems Planning Seminar on Energy Modelling, Washington, January 25-26, 1973.

6. Hoffman, K.C., A Systems Approach to Energy Resource Planning. Arbeitsseminar: Energiemodelle für die Bundesrepublik Deutschland, Jülich, 1975.

7. Lieb, W., Wirtschaftspolitische Massnahmen zur Einsparung von Energie im Verkehr. In: Energieeinsparung als neue Energiequelle (ed. K. Meyer-Abich), München, Hanser Verlag, 1979.

8. In this comparison the price differences between gasoline and diesel fuel can be neglected (problem of price building for coupled products in the oil industry). Other operating costs are assumed as similar.

9. Thoss, R., A Generalized Input-Output Model for Residual Management. Sixth International Conference on Input-Output Techniques, Vienna, April 22-26, 1974.

10. Döllekes, H.P., Ein Multisektorales Energie- und Umweltplanungemodell. Arbeitsseminar: Energiemodelle für die Bundesrepublik Deutschland, Jülich, 1975.

11. Thoss, R., Resolving Goal Conflicts in Regional Policy by Recursive Linear Programming. 13th European Congress of the Regional Science Association, August 28-31, 1973.

12. Hax, H., Entscheidungsmodelle in der Unternehmung. Hamburg-Reinbek, Rowohlt Verlag, 1974.

13. Compare the argumentation of Hoffman in /6/, p. 36.

14. See, e.g., Döllekes in /10/, p. 216.

15. Kapp, K.W., Zur praxis der Umweltpolitik und der Umweltplanung. In: Sozialisierung der Verluste (ed. K.W. Kapp), München, Hanser Verlag, 1972.

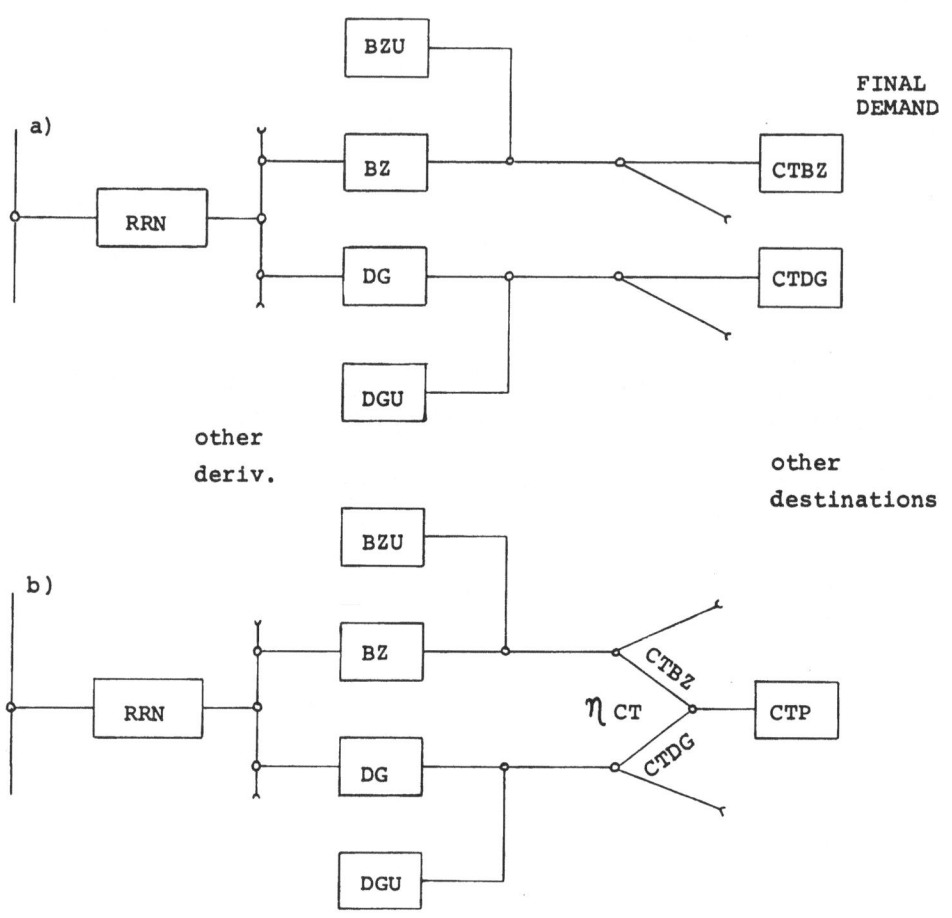

FINAL
DEMAND

other
deriv.

other
destinations

Terminology:

RRN    Index of crude oil refineries

BZ     Gasoline from domestic refineries

BZU    Imported gasoline

DG     Diesel fuel from domestic refineries

DGU    Imported diesel fuel

CTBZ   Gasoline demand for road transportation

CTDG   Diesel fuel demand for road transportation

CTP    Propulsion energy demand for road transportation

$\eta_{CT}$    Technical efficiency of converting fuel into propulsion energy
for road transportation

$-\!\!\prec$   Substitution and/or distribution nodes

Figure 1. Alternative designs of the road transportation in energy
systems modelling, demonstrated in the energy flow network
of H. Požar's energy optimization model /3/.

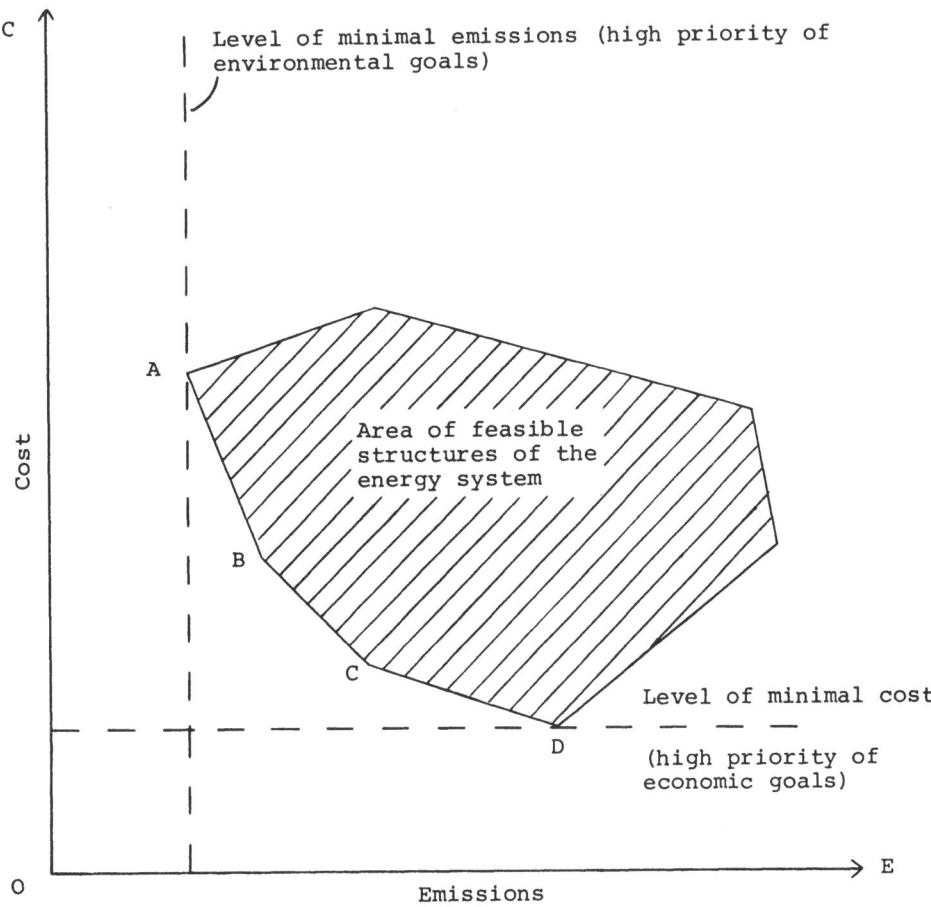

ABCD: Curve of pareto-optimal options for the structure of the
energy system

Figure 2a. Area of feasible solutions of the energy system
optimization problem, with regard to economic and
environmental aspects.

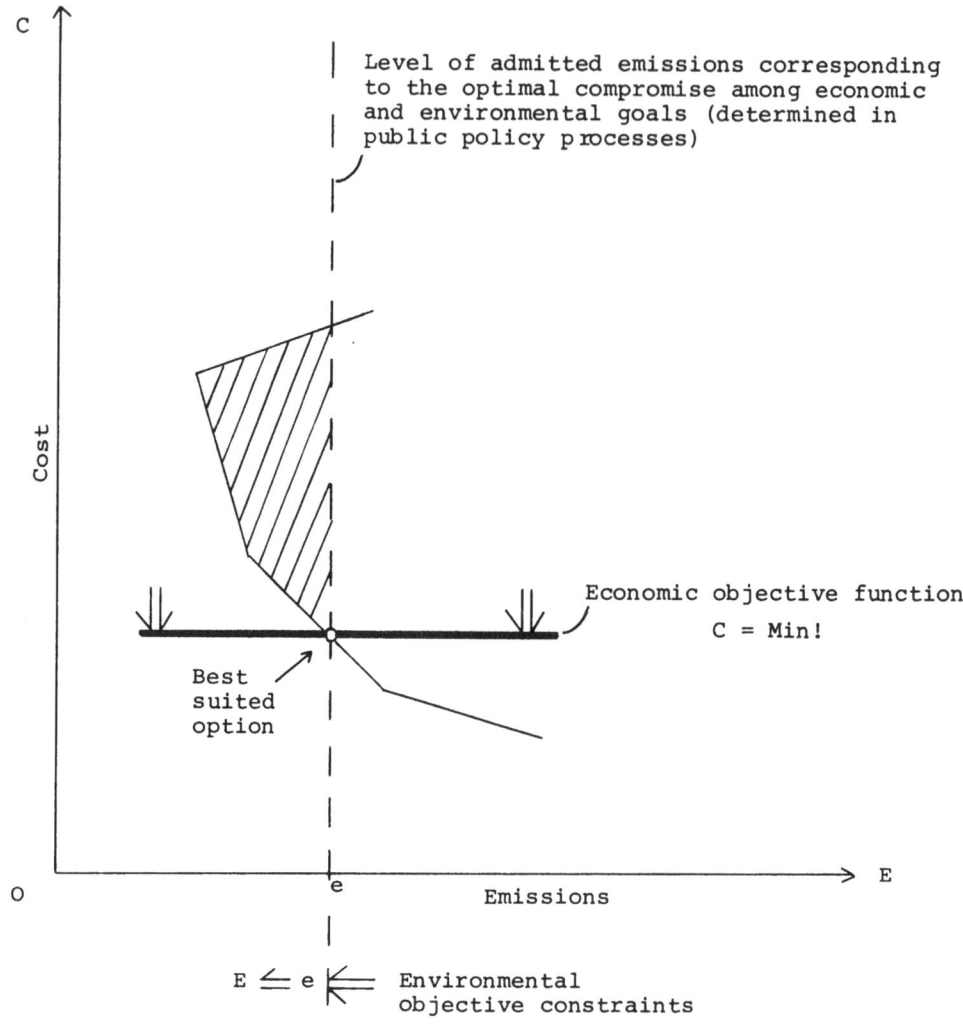

Figure 2b. Standard approach to energy system optimization with economic and environmental objectives.

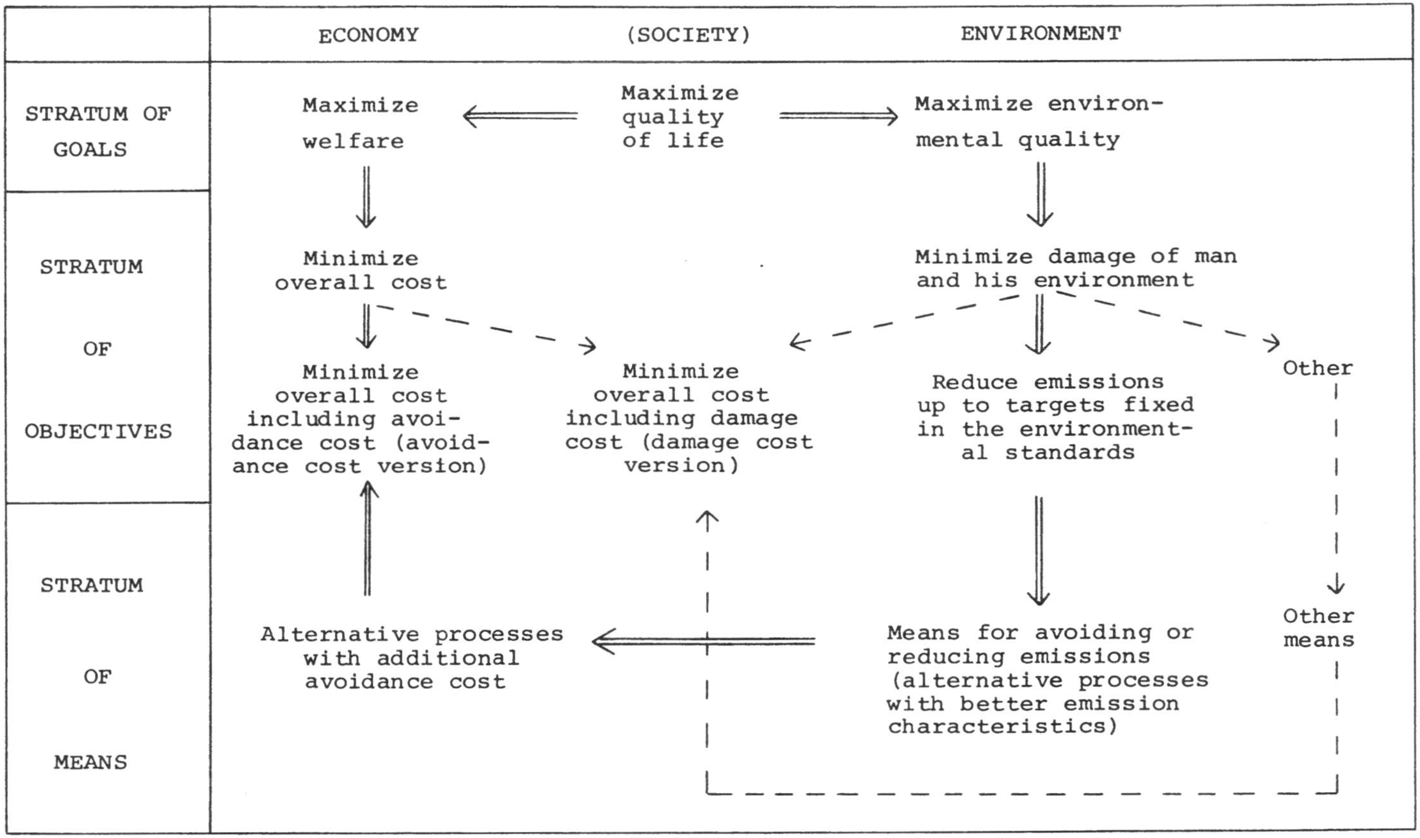

Figure 3. Interrelationship between economic and environmental goals and means in energy system optimization models.

# TREATMENT OF THE ARAB REGION IN GLOBAL MODELS

Nader Fergany

Arab Planning Institute

## Introduction

The purpose of this note is to discuss the treatment of the
Arab region in some of the better-known world models.  The main contention is that the region has been inadequately represented in global
models, especially from the vantage point of its alternative futures.
Therefore, future problems and trends derivable from these models for
the region do not represent a sound exploration of its future.

It is well known that a model is designed to serve a certain
specific aim and that it should not be used for purposes other than
that it purports to serve.  Hence, it can be claimed that global models
are not meant to tackle regional issues.  Nevertheless, models have
often been used for purposes other than what they were built to serve
even by the model builders.  In particular, global models have been
used to discuss regional futures and sometimes even to propose regional
strategies and policies.

Further, in as far as regional developments affect global
patterns and trends, inadequate representation of a region, in present
or potential conditions, implies inadequate representation of global
systems.

Most important, however, in the case of the Arab region is
that it passes through a critical epoch of its history in which a
clear vision of the alternative futures that could materialise is of
great value for sound decision-making.  This information base is not
provided by present global models.

In many ways, what we submit regarding the treatment of the
Arab region in global models applies equally well to other regions of
the third world.

## Significance of the Arab Region as an Entity, Past and Future

Although torn by strife and internal conflict, Arab countries
do form a socio-economic and political entity.

Unity derives from sharing the same language, and common cultural heritage and history.

The diversity of resources in the region provides for perfect complementarity among its countries. For example, some Arab countries have oil reserves that result in considerable financial resources, well beyond their capacity to absorb, but lack all other resources, most notably the local manpower needed to develop a viable socio-economic structure. Contrariwise, other Arab countries have either large populations providing for large work forces, or mostly unexploited vast agricultural resources, but lack the financial resources needed, among other things, for development.

From the point of view of the future of Arab countries, the consideration of the region as an entity takes on a decisive dimension. In spite of the financial affluence of some Arab countries, they are all underdeveloped and dependent societies. Further, contrary to a common misconception, the region as a whole is not rich in natural resources. The expectation is that it could remain food deficient. Even in energy sources the area could turn to be a net importer early in the next century. This situation fosters the urgent need to utilize the complementarity of the region´s resources to clutch on a process of genuine development.

While the present mix of resources in the region is favourable to development if used in an integrative way, it is not sufficient. In addition to natural resources, a society requires manpower and efficient utilization of human energies in development through effective institutions and social organization.

Fortunately, the Arab region is relatively rich in human resources. By the turn of the century, it is projected that the population of Arab countries will amount to about 300 million, larger than the population of the USA and a little less than that of the USSR. However, this represents only a potential that needs to be developed through appropriate socialization and social organization.

In summary, the process of development in the Arab region, and hence the boundaries of its future, hinge on two main determinants:

a) development of human resources, which requires structural socio-political change to arrive at an efficient form of social organization,

b) integration among Arab countries in a collaborative project of self-reliant development.

## Inadequate Representation of the Arab Region in Global Models

The inadequacy of representing the Arab region in global models stems from two interacting considerations. One is common to the representation of third world regions in global models in general, while the second relates to specifities of the region.

As we alluded earlier, the future of third world countries hinges on whether they engender a process of development entailing structural socio-economic and political change. This is contrary to the situation of industrialized countries where well-developed socio-economic and political systems have evolved and there may be no urgent desire, at least in many influential quarters, to change them structurally.

Various aspects of the process of development in the third world, however, require restructuring the socio-economic and political system. Foremost among these is human resources development aimed at maximizing the stock of knowledge, skills and creative energies of the people. No less important in this regard is the effective utilization of these resources to attain the goals of society. The critical factor in this process is social organization that permits the mobilization of human resources. It extends to leadership, ideology and means of popular participation. This formulation is at a high level of abstraction but it has implications on all other levels of the hierarchy of representation of the socio-economic system.

To give an example, education as a component of the process of development has to be enabling in the sense that it allows the individual to achieve his potential as a creative human being, prepares him for participating effectively in socio-economic and political activity and provides the foundation, among other things, for developing a local technological capability. Thinking of education in this way is a far cry from considering the proportion of illiterates or enrollment ratios in various stages of education which could mean quite different things in different societies. Such a conception also has implications for how the process of education is to be organized, in practice and in modelling, if certain developmental tasks, like the eradication of illiteracy, for example, are to be accomplished.

To generalize this example, we claim that adequate modelling of third world regions leading to meaningful exploration of their future should start from a clear and explicit conception of the process of development in such societies. This requirement is merely a restatement of the well-known dictum stipulating that a model has to have a logically consistent and coherent conceptual framework.

Hence, a model that represents industrialized and third world regions in the same basic structure does not allow for restructuring the socio-economic and political systems of the latter ones and, therefore, is insufficient to explore the alternative futures of these regions. Skipping structural socio-political change to their purely economic implications, expressed in terms suitable for industrialized economies, leaves out an essential aspect of modelling the process of development.

Some global modelling efforts arrive at the conclusion that the constraints to sustained growth in third world countries are socio-political and institutional in nature, nationally and internationally, but only after having conducted an exercise in modelling that did not involve these considerations. A notable example in this connection is the UN-sponsored study of the future of the world economy.

Embedding socio-political structural change in a model might entail adopting a certain political point of view. But it does not have to since, at least in theory, it is possible to consider more than one point of view in a modelling exercise. At any rate, the essential feature of a model that captures the essence of the future of less developed countries is that it cannot afford to be apolitical. Attempts at developing "unbiased" technical tools tended to adopt representations conforming to the socio-economic and political status quo reflecting conditions, and interests, in industrialized societies.

The one notable exception to this concept of "unbiased" models has been the Bariloche model of the world which adopted an explicit political point of view of the socio-economic conditions on the national and international levels, proposed a certain form of socio-political structural change, and incorporated it in a mathematical model.

The OECD study, Interfutures, gave adequate consideration to socio-political change within and among industrialized countries but not within and among third world countries, limiting itself to the nature of the relationship between the North and the South.

These varying approaches clearly reflect the interests and concerns of the modellers, which is understandable. But the point remains that the main developmental concerns of third world regions in general, and of the Arab region in particular, have not been reflected adequately in global models, with the exception of the Latin American world model.

It is to be granted that modelling, especially of the "hard" category, of structural socio-economic and political change poses considerable conceptual, analytical, size, and data problems. But there

only lies the challenge of arriving at an adequate representation of
third world regions, especially from the point of view of investigating
alternative futures.

We now turn to specific features of the treatment of the Arab
region in global models.

In most of the global models the Arab region was considered
mainly from the point of view of the oil reserves it contains. In
"Mankind at the Turning Point" for example, the region was given prom-
inence in the context of the analysis of the global energy situation.
The recommendations there included that the optimum price of oil would
be about 50 per cent higher than the 1974 level (!). The strategy
proposed for the region was to cooperate in securing oil supplies to
industrialized countries. This strategy was deemed optimum for it
leads to the maximization of financial accumulation of the region in
the industrialized countries (and not in accumulation conducive to
development in the region). This example, in addition, serves to illustrate points made earlier in this note.

Contrary to socio-economic and political realities, the Arab
regions were never treated as an entity in global models that undertook
an appropriately fine subdivision of the world. Non-Arab countries
were added and/or some Arab countries deleted in the formation of the
region(s) intersecting most with the set of Arab countries.

It was indicated earlier that a true process of development
in the Arab region would be an integrative venture. This was not a
consideration in any of the global models. Actually, in one of them,
the UN-sponsored model of the world economy, Arab countries were
divided between two regions in a subdivision of the world into 15
regions. Although trade relations were present among the regions of
the system, these do not constitute sufficient integrative linkages
between two less-developed socio-economic systems. It can be said that
this representation aborts the potential for a genuine process of
development in the region. In this sphere, mechanisms for socio-
economic integration among underdeveloped societies, that reinforce
the joint developmental efforts, need to be identified and properly
modelled.

Once more this classification of Arab countries in two regions
reflected the importance of the region, to industrialized countries, as
a reservoir of oil. For oil-rich Arab countries were grouped with oil-
rich non-Arab countries in the Middle East and Africa in one region.
Similarly, non-oil-rich Arab countries were grouped with non-oil-rich
non-Arab countries in Africa in another region. This, of course, shows

that oil endowment was the decisive factor in the classification.

In the Interfutures study, the "hard" modelling core of the work implied continuation of the present structural socio-economic and political conditions on the world level. But the qualitative analysis complementing it brought out the cultural homogeneity as well as the diversity and complementarity of resource endowment of the region. It also pointed out the great potential for cooperation among the countries of the region.

## Conclusions

The Arab region has been considered in global models in the context of the concerns and interests of organizations sponsoring, and individuals developing, the models. The region enjoyed special attention due to its huge oil reserves. Representation of the region was mostly in terms of continuation of socio-economic and political structures on the regional and international levels.

In consequence, global models have not treated the Arab region as a socio-economic and political entity. Further, the range of alternative futures available to the region as a result of restructuring socio-economic and political conditions, was not analyzed.

This, of course, can only be done in a truly regional model that starts from the region, allows for its specific features and incorporates the determinants of its future. There was an attempt to do exactly that a few years ago but it did not materialize. Currently, a similar effort is gaining momentum.

# IIASA´S ROLE IN GLOBAL MODELING

Gerhart Bruckmann

IIASA, A-2361 Laxenburg, Austria

The International Institute for Applied Systems Analysis is
an international research institute, founded in October 1972 at the
initiative of the academies of science or equivalent institutions in
12 countries, both East and West.  Since then, scientific bodies from
five more countries have joined the Institute, thus bringing the number of "National Member Organizations" (NMOs) to 17.  These scientific
institutions have set IIASA the task of tackling the complex problems
facing mankind today, problems that are the consequence of the success
of modern science and technology and that now require the joint efforts
of East and West in order to find approaches adequate to the challenge
of the future.  The Institute therefore conducts and stimulates research
on problems of modern societies having international importance.

With these goals in mind, IIASA, in its early days, was also
faced with the question to what extent it should involve itself in
"Global Modeling".  When IIASA was founded, "Limits to Growth" had
just been published.  This first "Global Model" had not only created
worldwide discussion and criticism, but had also given rise to what
is - in retrospect - being referred to as a "Global Modeling Movement",
attempts by many other institutions to "do better".

The first global model to be developed after "Limits to
Growth", the Mesarovic/Pestel model, was much larger in size and in
manpower requirements.  If IIASA had involved itself in elaborating
its own global model, this venture might have absorbed too high a part
of its entire capacity; if, however, IIASA would not involve itself in
global modeling at all, there was danger of bypassing an important new
development of unknown potential.

In this situation, IIASA decided to take up a "monitoring
role": to serve as a focussing point for an exchange of opinions and
findings, to help avoid overlapping and duplication of efforts, and to
follow closely to what extent the findings could be of use to other
IIASA work.

This goal was implemented mainly by conducting a series of

global modeling conferences. Whenever a major model reached a state of completion IIASA invited its authors to present the model before an international community of experts; the advantage on the side of the authors being that the results of the discussion could still enter the work before its final publication.

It was obvious that the First Global Modeling Conference, held in April/May 1974 was devoted to the Mesarovic/Pestel model. How much importance the authors attached to this conference can be judged from the fact that the (five) technical volumes of the model appeared as proceedings of this conference[1]. The popular volume, "Mankind at the Turning Point"[2] was published half a year after the conference.

At each IIASA global modeling conference, the last day was devoted to a presentation and discussion of other global modeling work. This did not only serve for a mutual information within the rapidly growing "global modeling community", but also facilitated the selection of the next model to be invited for an IIASA conference. In this sense, the model elaborated by a group of scientists of the Fundacion Bariloche (Argentina) was invited to be the main model of the Second IIASA Global Modeling Conference (October 1974). The presentation of this model met vivid interest; it had been set up as a normative counter model against both the Forrester/Meadows and the Mesarovic/Pestel model. Its purpose was not to shed light on the future of the globe but rather to assess paths which would secure a decent standard of living to the poorer parts of this world. The conference proceedings[3] were published long before the final version of the model[4].

The main model to be presented and discussed at the Third IIASA Conference (September 1975) was MOIRA, worked out by a Dutch team under the chairmanship of Hans Linnemann. The main purpose of this model was an investigation of ways and means to reduce hunger in the world. The model, therefore, was very elaborate and detailed in the food and agriculture sector, but deliberately treated the rest of the economy as more or less exogenous. Again, the conference proceedings appeared long before the book[5], [6].

The Fourth Conference (September 1976) was devoted to a presentation and comparison of two models, of the SARUM model developed by the Department of the Environment (United Kingdom) and the MRI model developed by the Polish Academy of Sciences. The confrontation of a Western type and an Eastern type model proved extremely fruitful. Whereas SARUM was published in full elsewhere[7], the only full version of MRI (outside of Poland) is contained in the conference proceedings[8].

The Fifth Global Modeling Conference (September 1977) was not so much devoted to a single model, but to an approach, i.e.,

"Input-Output Approaches in Global Modeling". Amongst the models
presented were the Leontief model[9], the Japanese model FUGI, the Swiss
model ZENCAP (Fritsch/Codoni/Saugy)[10], and the model by Bottomley. At
this conference, for the first time in an IIASA global modeling conference, a report was also given on modeling work in the Soviet Union.
The conference proceedings contain the only full version of FUGI published in any language other than Japanese.

The Sixth Global Modeling Conference (September 1978) deviated
in scope and format from all earlier conferences. Its purpose was to
assess the state of the art. With that goal in mind, a questionnaire
was designed and sent out to all major global models; the replies to
the questionnaire formed the background material of the conference.
The conference itself was not organized by model, but by question. All
main results of the conference will be contained in a forthcoming
book[12]; other papers presented at the conference are available from
IIASA upon request.

The Seventh IIASA Global Modeling Conference (September 1979)
was devoted to a special topic, "Environmental Aspects in Global Modeling". The conference brought together environmentalists on one side
and global modelers on the other side. It definitely served to increase
mutual understanding of the difficulties each side encounters. The
proceedings are still in print[13].

The Eighth Global Modeling Conference (July 1980) was devoted
to a comparison of "International Economic Modeling" work. Econometric,
input-output, and general equilibrium models were presented and discussed.

The Ninth IIASA Global Modeling Conference, to be held
September 14-18, 1981, will be different again; whereas all earlier
conferences were inside-oriented (discussions amongst the global modeling community), the ninth conference will be outside-oriented. Under
the working title "Global Modeling at the Service of the Decision
Maker", a wide audience of policy makers, government advisors, and
media experts shall be informed about global modeling. Similarly as
at the sixth conference, all major models will be asked a set of questions of particular importance to policy makers. They will work out
the answers before the conference; these answers will be sent to participants as background material. The conference itself will be focussed around these questions; in each session, one of these topics will
be taken up and the answers which the different models give will be
confronted and compared. It will definitely be a unique opportunity
to have all major models together at one conference, giving their
opinions on issues vital to mankind.

Summarizing, it may safely be stated that IIASA has been able to actually implement the self-chosen monitoring role in the field of global modeling.

References

1. Mesarovic, M.D. and Pestel, E. (ed.): Multilevel Computer Model of World Development System. Proceedings of the IIASA Symposium, Vol. I-VI. SP-74-1/6. Vienna 1974.

2. Mesarovic, M. and Pestel, E.: Mankind at the Turning Point (New York, Dutton, 1974).

3. Bruckmann, G. (ed.): Latin American World Model. Proceedings of the Second IIASA Symposium on Global Modeling. IIASA CP-76-8. Vienna 1976.

4. Herrera, A.D., Skolnik, H. et al.: Catastrophe or New Society? A Latin-American World Model, Ottawa, International Development Research Centre, 1976.

5. Bruckmann, G. (ed.): MOIRA - Food and Agriculture Model. Proceedings of the Third IIASA Symposium on Global Modeling. IIASA CP-77-1. Vienna 1977.

6. Linnemann, H. et al.: MOIRA - Model of International Relations in Agriculture. North Holland Publishing Co., Amsterdam 1979.

7. Bruckmann, G. (ed.): SARUM and MRI. Description and Comparison of a World Model and a National Model. IIASA Proceedings Series, Vol. 2. Pergamon Press 1979.

8. Roberts, P.C. et al.: SARUM 76 - Global Modelling Project. Research Report 19. UK Department of Env. and Transport. London 1977.

9. Leontief, W. et al.: The Future of the World Economy. A United Nations Study. Oxford University Press. New York 1977.

10. Codoni, R. and Fritsch, B.: Project ZENCAP - Capital Requirements of Alternative Energy Strategies. Institut f. Wirtschaftsforschung der ETH. Zurich 1980.

11. Bruckmann, G. (ed.): Input-Output Approaches in Global Modeling. IIASA Proceedings Series, Vol. 9. Pergamon Press 1980.

12. Meadows, D., Richardson, J., Bruckmann, G. (ed.): Groping in the Dark. John Wiley and Sons (in preparation).

13. Bruckmann, G. (ed.): Environmental Aspects in Global Modeling. IIASA Proceedings Series. Pergamon Press (in preparation).

# QUASI-MODELS OF PRICE EVOLUTION AND THEIR QUALITATIVE PROPERTIES

Vladimir B. Bajić[*], Bratislav J. Petrović[**]

[*]PTT Educational Center, Belgrade, Yugoslavia
[**]Faculty of Organizational Sciences, Belgrade, Yugoslavia

## 1. Introduction

There is great difficulty in estimating the models of time dependent prices in a large market system as a consequence of the large number of goods. The estimated model will have a large number of equations. The parameters´ estimation and analysis of such models are almost impossible in practice. A simple technique is used in order to overcome these problems. This technique is based on an aggregation procedure which is not so restrictive as that proposed by Lange /1/, but neither as flexible as general decomposition - the aggregation method introduce d by Šiljak /2/, /3/. The model obtained by the proposed aggregation technique is denoted as quasi-model because it governs the time evolution of quasi-prices which are specially constructed scalar functions of real prices preserving their trends of changes. The resulting quasi-models may be deterministic or stochastic. The qualitative properties of such models will be considered.

The control actions, denoted as fundamental control, which essentially change the evolution of quasi-prices are introduced into the quasi-model and can be implemented into the price evolution mechanism.

## 2. Quasi-Prices

Let us consider k commodities and/or services (goods in further text) which appear in a large market system. Suppose that in this market system the past evolution of prices for these goods is known on time interval $[t_b, t_1]$. Here, $t_1$ is the present moment when we begin our consideration. Now, to acquire some knowledge of the price evolution process in the future on the basis of this information, one can apply some of the identification methods. In the models obtained, the number of equations will be enormously large and, thus, without great practical value.

For this reason, we will consider an approach to reduce the original large number equations. Let us now make an aggregation of original k goods into n composite goods. The aggregation will be based upon the assumptions:

1) that the change of prices of all subgoods in the composite good do not have the opposite direction and that this holds true in the near future; or

2) that the change of difference among subgoods prices in the composite good and the price of some reference good, do not have the opposite direction and that this holds true in the near future.

With $G_i$, $i = 1,2, \ldots,n$ we denote i-th composite good, consisting of $n_i$ subgoods, the prices of which are $p_{ji}$, $j = 1,2, \ldots,n_i$. We now define for each composite good $G_i$, $i = 1,2, \ldots,n$ the subsidiary scalar function $q_i(p_i)$, where $p_i$ is the price vector of subgoods in the considered composite good. The functions $q_i(p_i)$ should be constructed to directly follow the change of components $p_{ji}$ of vector $p_i$, i.e., if any of $p_{ji}$, $j = 1, \ldots,n_i$ increase (decrease) the function $q_i(p_i)$ does the same. In the second case, quantities $p_{ji}$ will denote the difference between the price of the subgood and the price of the reference good.

Because of the way they are constructed, the functions $q_i(p_i)$ present some "integral" price for all subgoods in each composite good and follow the change of the original prices of subgoods. Thus, we call them quasi-prices. Through the quasi-prices one can observe global trends in the time evolution of real prices.

If the values of prices $p_{ji}$, $j = 1, \ldots,n_i$, $i = 1, \ldots,n$ are known in the time interval $\left[t_b, t_1\right]$ and if functions $q_i(p_i)$ are chosen, then the values of quasi-prices in the same time interval are also known.

Now, instead of looking for models which describe real price evolution process we can use identification methods to determine models describing the trends of time evolution of quasi-prices and we may consider quasi-prices as state variables of the new model. This model we denote as quasi-model.

The above aggregation procedure contains as a special case the aggregation method proposed by Lange /1/, where subgoods prices in one composite good "vary always in the same proportion". Also, the quasi-prices of composite goods, in particular, may be a linear combination of subgoods prices as in /1/, but it is not necessary. This technique may considerably reduce the number of equations. Although this reduction involves inaccuracies, the advantage of quasi-models reduced order makes their application acceptable. We also assume that the identified quasi-models are in the form of differential equations system.

## 2.1. Deterministic Quasi-Model

The deterministic quasi-model can be described by the equations of the form

$$\dot{q} = f ( q ) \tag{1}$$

where $q(t) \varepsilon R^n$ is quasi-price vector, $t \varepsilon T$, $T$ is the time interval $T = \{t: a < t < +\infty\}$ where a is the real number of symbol $-\infty$, and $f(.)$ is the vector function $f: R^n \rightarrow R^n$. Generally speaking, there are no constraints on the function $f(q)$ except its smoothness in order to ensure existence of the solutions $q(t; t_o, q_o)$ of (1), where initial conditions $(t_o, q_o) \varepsilon T \times R^n$. We will restrict our consideration to the case when system (1) can be represented in the form

$$\dot{q} = A(q)q + b(q) \tag{2}$$

where $n \times n$ matrix A is continuous as well as vector b. The form (2) of equations (1) is suitable for the analysis which follows in the next sections. In the case of competitive equilibrium, the right side of equation (2) reflects properties of the subgoods excess demand functions.

To consider connective properties of the system (2), we write the elements $a_{ij}$ of matrix A in the form

$$a_{ij}(q) = \begin{cases} -z_i(q) + \bar{e}_{ii} z_{ii}(q) \ , & i = j \\ \\ \bar{e}_{ij} z_{ij}(q) & , & i \neq j \end{cases} \tag{3}$$

and vector b elements $b_i$ in equs. (2) in the form

$$b_i = \bar{l}_i w_i(q) \ , \qquad i = 1, \ldots, n \tag{4}$$

Elements $\bar{e}_{ij}$ and $\bar{l}_i$ are

$$\bar{e}_{ij} = \begin{cases} 1 \ , & \text{if } z_{ij}(q) \not\equiv 0 \\ \\ 0 \ , & \text{if } z_{ij}(q) \not\equiv 0 \end{cases} \tag{5}$$

$$\bar{1}_i = \begin{cases} 1, & \text{if } w_i(q) \not\equiv 0 \\ \\ 0, & \text{if } w_i(q) \equiv 0 \end{cases} \qquad (6)$$

Elements $\bar{e}_{ij}$ define n x n fundamental interconnection matrix $\bar{E} = (\bar{e}_{ij})$ /3/, and elements $\bar{1}_i$ define fundamental interconnection vector $\bar{1} = (\bar{1}_i)$ , /2/.

In order to study interaction influences we will replace every unit element $\bar{e}_{ij}$ and $\bar{1}_i$ with time functions $e_{ij}(t)$ and $1_i(t)$, respectively, such that

$$e_{ij} : T \rightarrow [0,1] \quad , \quad \forall t \varepsilon T, \quad e_{ij}(t) \varepsilon C^o(T)$$

$$\qquad (7)$$

$$1_i : T \rightarrow [0,1] \quad , \quad \forall t \varepsilon T, \quad 1_i(t) \varepsilon C^o(T)$$

Time-varying functions $e_{ij}(t)$ and $1_i(t)$ define n x n time-varying inter-connection matrix $E(t)$ and time-varying n-dimensional vector $1(t)$ res-pectively. In matrix $E(t)$, i.e., vector $1(t)$ elements $\bar{e}_{ij} \equiv 0$, i.e., $\bar{1}_i \equiv 0$ are on the same position as in $\bar{E}$, i.e., $\bar{1}$. Due to the fact that $E(t)$ /1(t)/ is derived from $\bar{E}$ /$\bar{1}$/ we write $E(t) \varepsilon S_{\bar{E}}$ /1(t) $\varepsilon S_{\bar{1}}$/.

If in the model (2) we intend to implement the influence of coupling intensity between $\dot{q}_i(t)$ and $q_j(t)$ then we replace matrix $\bar{E}$ with $E(t) \varepsilon S_{\bar{E}}$ and in (3) corresponding elements $\bar{e}_{ij}$ become time-varying functions $e_{ij}(t)$. The same procedure should be done for the term b in the equats. (2).

## 2.2. Stochastic Quasi-Model

By introducing interconnection matrix $E(t)$ and interconnec-tion vector $1(t)$ in the model (2), its inaccuracy is reduced. To go a step further in this direction we shall derive the model of quasi-price time evolution process with stochastic disturbances. In this sense, let us consider two competitive industries $J_1$ and $J_2$ each one offering some composite goods on the same market system. With the same identi-fication procedure, model (2) will be the same for both industries. In general, for each of the considered industries, fluctuation of the interactions between some composite goods quasi-prices are not under their direct control and it is reasonable to decompose matrix A in the following way.

$$A = A_1(q) + B_1(q) \text{ for industry } J_1 \qquad (8)$$

and

$$A = A_2(q) + B_2(q) \text{ for industry } J_2 \qquad (9)$$

Let us consider only eq. (8). In the matrix $B_1$ all terms of A which do not present such directly controlled interactions are included. This implies the relativity of such decomposition. In this way from (2) and (8) /(9)/ we get the model form

$$dq = A(q)q \ dt + b(q) \ dt + B(q)q \ dt \tag{10}$$

where $z(t) \varepsilon R$ is a normalised Wiener process with

$$\xi( \ z(t_1) \ - \ z(t_2) \ )^2 \ = \ |t_1 - t_2| \tag{11}$$

where $\xi$ is mathematical expectation. Matrix B describes the influence of stochastic disturbances on the quasi-price evolution process. We assume that the elements $a_{ij}$ of Matrix A and $b_i$ of vector b in (10) will be of the form (3) and (4) while elements $b_{ij}$ of matrix B will be written in the form:

$$b_{ij} = \bar{l}_{ij}\dot{x}_{ij}(q) \tag{12}$$

where the elements $\bar{l}_{ij}$ are defined as

$$\bar{l}_{ij} = \begin{cases} 1 \ , & x_{ij}(q) \neq 0 \\ \\ 0 \ , & x_{ij}(q) = 0 \end{cases} \tag{13}$$

and the fundamental interconnection matrix $\bar{L} = (\bar{l}_{ij})$ is obtained.

Further reduction of inaccuracies of model (10) can be done by replacing matrix $\bar{L}$ with interconnection matrix $L(t) = ( \ l_{ij}(t) \ )$ where

$$l_{ij}(t) = \begin{cases} l_{ij}(t) \ , & \text{if } \bar{l}_{ij} = 1 \\ \\ 0 \ , & \text{if } \bar{l}_{ij} = 0 \end{cases} \tag{14}$$

where functions $l_{ij} : T \to [0,1]$ are $l_{ij}(t) \varepsilon C^o(T)$. If L(t) is derived from $\bar{L}$ we denote it as $L(t) \varepsilon S_{\bar{L}}$.

The stochastic part in (10) is introduced to describe more realistically the random fluctuations in q, occuring due to unpredictable change in technology, population, tastes of consumers, etc., and because of random perturbations of an internal mechanism in the quasi-price time evolution process. The functions $l_{ij}(t)$ measure the stochastic disturbances influence in the market system. The qualitative analysis of the model having the form (10) for b ≡ 0 was considered in /2/, /3/.

## 3. The Influences of Interactions

We will turn our attention to two particular problems concerning deterministic and stochastic quasi-models. First, we shall consider the deterministic quasi-model of the form (2) and state questions as follows:

1) Is it possible, when $q(t,t_o,q_o)$ is unbounded on $T_o$, where $T_o = \{t : t_o \leqslant t < +\infty\}$, $t_o \in T$, to make the quasi-price evolution process in the considered market system bounded, by excluding some composite goods?

2) If the quasi-price evolution process in the considered market system is unbounded on $T_o$, is it possible to make it bounded by introducing new composite goods in the market?

To answer these questions some explanations are necessary. When k - th composite good disappears (is excluded) from the market then

$$e_{ik} = e_{kj} = 0 \quad , \qquad i,j = 1,2, \ldots,n \qquad (15)$$

When several composite goods are excluded from the market we do the same. In fact, from the original fundamental interconnection matrix $\bar{E}$ we derive interconnection matrix $\bar{E}_o$, having the same order as $\bar{E}$. Matrix $\bar{E}_o$ is obtained by substituting all elements $e_{ik}$ and $e_{kj}$, $i,j = 1,2, \ldots,n$ with zero for all k which are concerned with excluded composite goods.

In further consideration, real numbers $\alpha_i > 0$ and $\alpha_{ij} \geqslant 0$ are such that $\alpha_i > \alpha_{ii}$, and real numbers $\beta_i$ are nonnegative. We shall set different constraints on the elements of matrix A and vector b which will not all be used at the same time. These constraints are

A) $z_i(q)|q_i| \geqslant \alpha_i g_i(|q_i|)$, $z_{ij}(q)q_j \leqslant \alpha_{ij} g_j(|q_j|)$

$\quad$ $i,j = 1,2, \ldots,n; \forall q \in R^n$

where $g_i(x)$ are well known comparison functions of class K /4/;

B) $z_i(q) \geqslant \alpha_i$ $\quad$ , $\quad$ $|z_{ij}(q)q_j| \leqslant \alpha_{ij}|q_j|$

$\quad$ $i,j = 1,2, \ldots,n; \quad \forall q \in R^n$

C) $|w_i(q)| \leqslant \beta_i$ $\quad$ , $\quad$ $i = 1,2, \ldots,n; \forall q \in R^n$

D) $b_i(q) \equiv 0$ $\quad$ , $\quad$ $i = 1,2, \ldots,n; \forall q \in R^n$

Also, we define the matrix $\bar{A} = (\bar{a}_{ij})$ so that

$$
\bar{a}_{ij} = \begin{cases} - \alpha_i + \bar{e}_{ii} \alpha_{ii} & i = j \\ \\ \bar{e}_{ij} \alpha_{ij} & i \neq j \end{cases} \tag{16}
$$

where $\bar{e}_{ij}$ are elements of the fundamental interconnection matrix $\bar{E}$.

If r composite goods are excluded from the market system, the original fundamental interconnection matrix is changed into interconnection matrix $\bar{E}_r$. The corresponding matrix $\bar{A}$ defined by eq. (16) is also transformed into matrix $\bar{A}_r$.

Now, we have the following statement.

Proposition 1 - When the strongest interactions occur in the market system then, if the quasi-price evolution process described by (2) produces unbounded quasi-prices and

i) if A) and C) hold and if, by excluding r composite goods, the corresponding matrix $\bar{A}_r$ is Metzlerian and Hicksian, then the corresponding system is connectively uniformly bounded in the large; or

ii) if B) and C) hold and $\bar{A}_r$ is Metzlerian and Hicksian, then the corresponding system is connectively exponentially and ultimately bounded in the large; or

iii) if A) and D) hold and $\bar{A}_r$ is Metzlerian and Hicksian, then the corresponding system is connectively asymptotically stable in the large; or

iv) if B) and D) hold and $\bar{A}_r$ is Metzlerian and Hicksian, then the corresponding system is connectively exponentially absolutely stable.

The proof of the proposition can be performed on the basis of /2/ and /3/. It should be noted that in these cases it must be possible to exclude composite goods from the market system.

Now, it is assumed that by introducing new composite goods into the market system, the resulting quasi-price evolution process is governed by linear model

$$
\dot{q} = A q + b \tag{17}
$$

where A is constant $n_o \times n_o$ Metzler matrix, $n_o > n$, and q is quasi-prices vector with the extended dimension $n_o$. The vector b is constant. For the system (17) $n_o \times n_o$ fundamental interconnection matrix $\bar{E}$ is derived from the original one by taking into account the new interactions occurring when new composite goods are introduced. Then we have:

Proposition 2 - If the market system produces quasi-prices unbounded on T and new composite goods are introduced into that market system so that resulting evolution of quasi-prices is governed by (17),

the quasi-prices are still unbounded. Proof: Because of the Metzlerian
property, matrix $\bar{A}$ implies /2/ that solution $q(t, t_o, q_o)$ of (17) with
interconnection matrix $E(t) \in S_{\bar{E}}$ is always under the solution
$\bar{q}(t, t_o, \bar{q}_o)$ of (17) with the strongest interactions, provided $q_o \leq \bar{q}_o$
(inequalities are elementwise), then it immediately follows that
Proposition 2 is true.

In the case of the stochastic system we need the following
constraints on matrix A and B:

E) $z_i(q) \geq \alpha_i$, $\quad |z_{ij}(q)| \leq \alpha_{ij}$, $\qquad i, j = 1, 2, \ldots, n$; $\forall q \in R^n$

F) $|x_{ij}(q)| \leq \mathcal{Y}_{ij}$, $\quad i, j = 1, 2, \ldots, n$; $\forall q \in R^n$

where $\mathcal{Y}_{ij} \geq 0$ are real numbers.

We also define the matrix $\bar{C}$ as

$$\bar{C} = \bar{A} + \bar{A}^T + \bar{B} \tag{18}$$

where matrix $\bar{B}$ is defined as $\bar{B} = (\bar{b}_{ij})$ with

$$\bar{b}_{ij} = \bar{1}_{ij} \ \text{ij} \ \sum_{k=1}^{n} \bar{1}_{ik} \ \mathcal{Y}_{ik} \tag{19}$$

If r composite goods are excluded from the market system we obtain
matrix $\bar{C}_r$ in an analogous way as matrix $\bar{E}_r$, i.e., $\bar{A}_r$. Now the following proposition holds:

Proposition 3 - If the time evolution process described by
(10) produces process $q(t; t_o, q_o)$ stochastically unstable in the mean,
when the strongest interactions occur in the market system, then if
E), F) and D) hold and if, by excluding r composite goods, the corresponding matrix $\bar{C}_r$ is negative dominant diagonal then the equilibrium
process of the corresponding system is globally exponentially connectively stable in the mean.

Proof can be performed on the basis of /3/.


4. Quasi-Models in the Presence of Fundamental Control

Let us assume that when all subgoods in one composite good
originate from the same industry or source then the quasi-price of
this composite good can be controlled by this industry or source. In
such a situation it is possible that the quasi-price be raised so high
that no transactions of subgoods can occur. On the other hand, it can
be decreased abnormally. So, when considered actions appear on the
market system there is no competitive market economy.

In order to form a model of the time development of quasi-
prices in this situation, we shall start from the model of the form (2)

$$\dot{q} = Aq + b \tag{20}$$

which is obtained when there are no previously considered actions.

We present the system of perturbed motion, derived from (20), as

$$\dot{y} = Ay \tag{21}$$

Further, it is assumed that system (21) is asymptotically stable (i.e., matrix A is Hurvitz matrix). Now, if actions of considered type appear in the market system, we still might have an asymptotically stable system. This essentially means that the rates of changes of y, i.e., q are changed with these actions. The previous conclusion concerning system (21) excludes additive terms in (21) and also indicates that actually matrix A is changed. We denote these actions as control inputs in vector form u(t) and we get A = A(u(t)). The simple form of this functional dependence is

$$A(u(t)) = A + U(t)M(t) \tag{22}$$

where $u(t) \in R^r$, $r \leqslant n$, is the input vector, matrix $U: R^r \to R^{n \times nr}$ is

$$U(t) = \text{diag} ( u^T(t), \ldots, u^T(t) ) \tag{23}$$

and matrix M is $M: T \to R^{nr \times r}$. Now, system (21) becomes

$$\dot{y} = ( A + U(t)M(t) )y = A_o ( u(t) )y \tag{24}$$

which is the classical homogenous bilinear system /5/, /6/.

As can be seen from (24), inputs u(t) influence the essential properties of matrix $A_o$ and for that reason we call them fundamental control.

For the case of the nonlinear stochastic model of type (10), matrix A should be substituted by

$$A_o(u) = A + U(t)M(t) \tag{25}$$

The relation (25) gives us the possibility for computing fundamental control u(t) so that the complete system obtains the desired properties.

## 5. Conclusions

In this paper we considered a simple technique for modeling a large-scale market system. Quasi-model approach, used in the paper, may considerably reduce the order of models obtained.

Because of the aggregation applied, inaccuracies occur in the model. In order to reduce these inaccuracies, time-varying interactions of deterministic and stochastic types were introduced.

For the deterministic case, conditions are derived, which

ensure that the market system producing unbounded quasi-prices in the case of strongest interactions, can be stabilized by excluding some composite goods from the market. Similar results are obtained for the stochastic case. Also, it was shown that under proper conditions the market system that produces unbounded quasi-prices cannot be stabilized by introducing new composite goods into the market system.

Models are also derived in the case when in the market system actions exist consisting of changing the quasi-prices and these actions were included in the model.

## References

1. O. Lange, "The Stability of Economic Equilibrium", in Reading In Mathematical Economics, part I, P. Newman, ed., Baltimore, Md: 1968.

2. D.D. Šiljak, "Competitive Economic System: Stability, Decomposition and Aggregation", IEEE, Trans. On. Aut. Control, Vol. AC-21, No. 2, April 1976.

3. D.D. Šiljak, "Large-Scale Dynamic Systems - Stability and Structure", North-Holland, New York, 1978.

4. W. Hahn, "Stability of Motion", Springer-Verlag, Berlin, 1967.

5. C. Bruno, et al., "Bilinear Systems: An Appealing Class of 'Nearly Linear' Systems in Theory and Applications", IEEE, Trans. On. Aut. Contr., Vol. AC-19, No., August 1974.

6. R.R. Mohler, "Bilinear Control Processes", Academic Press, New York, London, 1973.

# A MODEL OF REGIONAL INTERACTIONS
## CONSIDERING ENERGY DEFICITS

G.G. Pirogov

All-Union Institute of Systems Science
29 Ryleyev Str., Moscow 11934, USSR

Today every nation being drawn into the system of international labour division becomes a part of the world economy.  Scientific
and technological progress as well as further development of productive forces results in a more profound industrial specialization and
labour division (initially at intranational and later at international
levels) significantly increasing the influence of each national
economy over the world economy and vice versa.

The controversial process of integration and differentiation
running simultaneously in the modern world economy cannot be conceived
in isolation from acute problems facing mankind such as energy and raw
material supply, food provision, demographic explosion and others, constituting all together the so-called set of global problems /1, 2/.

Attempting to build a model of regional interactions (further
- the RIM) as a part of a system of global development models /3/, the
model builder at necessity encounters some difficulties.  The RIM has
to meet the following requirements:

1. It should be simple and at the same time flexible enough
to allow for analysis of various alternative development scenarios on
the basis of different data sets;

2. It should be "open" in the sense that the values of some
variables are allowed to be scenario-specified;

3. It should be valid for a sufficiently long prediction
period.

In other words, the RIM, though simple, has at the same time
to possess a specification which is flexible and easy to interpret.

While developing the RIM, it is necessary to take explicitly
into consideration the dependency of the Gross Regional Product (GRP)
on mineral resource extraction costs.  This necessity arises from the
present state of the world economy with, firstly, the natural resources
deficit being one of the most significant limitations to economic

development and, secondly, the large scale at which interregional flows
of raw materials are influencing the pattern of modern international
trade.

We shall assume that in a one-sector model the volume of GRP
is dependent not only on labour and capital factors but also on an
aggregate national resource. This aggregate resource will be represented by the energy resource for it is one of the most important raw
materials in international trade and better provided with statistics
than any other resource. The energy resource will be explicitly introduced into the model in the form of primary energy resource, covering
coal, lignite, crude oil, natural gas liquids, hydropower and nuclear
energy. All these resources are measured in million tons of coal
equivalent.

We are nowadays facing a gradual depletion of national
resources in rich convenient for extraction deposits as well as a transfer to inferior deposits with less concentration and more extraction
and transportation difficulties. This transfer brings about a continuous rise in extraction costs and raw material prices. It is reasonable, therefore, to introduce a resource depletion function into the
RIM related to the extraction cost-resource price increase rate. This
relationship could be expressed by subtracting a certain amount from
the GDR. This way the resource depletion-rise in resource price effect
cuts the actual value of the GDR compared with the potential one which
would be attainable in the absence of this effect and the resource depletion is modelled as a gradual and continuous process.

This kind of a model is quite suitable for regions closed
from the view point of energy demand/supply, i.e., those regions where
extraction is basically conditioned by industrial extraction capacities
and meet domestic demand.

As the only resource treated by the RIM is the energy resource,
the interregional resource flows will be represented almost exclusively
by crude oil. This representation is in conformity with the existent
foreign trade pattern for this resource. Up to the 1970s (the beginning of the energy crisis) the supply and prices problems in the world
market were rather simple. The supply generally had satisfied demand
and the price was being kept at a monopolistic high level, related to
marginal deposits extraction costs and the costs of transportation.
The resource deposits were controlled to lesser or greater extent by
international monopolies doing their best to exploit them at maximum
rates /6/.

By the beginning of the seventies, and especially during the
seventies, the oil situation had changed drastically. Rich deposits

were nationalized almost everywhere. The Organization of Petroleum
Exporting Countries (OPEC) has emerged and gained strength. World
prices for oil started to rise. At the same time, inflation developed
at a fast rate in industrial capitalist countries.

In these circumstances, the gap between world oil prices and
regional extraction costs has widened. The OPEC countries, in view of
the existent world inflation, shifted their policies to oil conservation through production cuts. Today the major determinants of oil
production are not the world prices for oil but other factors.

Most of the rich deposits are not exploited at maximum rates.
In this situation, the concept of a marginal supply source has changed.
Oil deposits similar to those in Alaska or the North Sea are insufficient to cover the deficit resulting from underproduction in the Middle
East. For this reason they stop being marginal and do not determine
the level of prices in the world market. On the other hand, governmental policies in the leading capitalist industrial countries shifted
to deliberate limitation of oil imports with the aim of eliminating
difficulties in the balance of payments and establishing independency
from energy supplies from abroad. Therefore, on the demand side, the
world price is no longer the single factor determining imported quantities. Thus, at the present time, the influence of market factors on
energy resource flows is significantly weakened and the classical
supply-demand model is not applicable any more.

Under these conditions, alternative energy resources form
part of the marginal deposits. The costs of alternative energy supply
determine the limits towards which oil prices are driving. Setting
aside the distant future, when rather exotic energy sources would be
utilized, and coming back to the reality of the near future, the first
decade of the twenty-first century, we must think of coal and nuclear
energy (in various combinations) as the feasible alternative source of
energy supply. In order to make both of them play a significant role
in the energy balance formation, tremendous capital has to be invested
in their development. Moreover, the existent transportation system
(based on motor transport in general and on private automobiles in
particular) as well as the existent system of diffuse settlements in
industrial capitalist countries (especially in the USA) put forward
an urgent demand for liquid fuel as the main energy resource. The
nuclear energetic would hardly provide motor transport with the fuel
needed. For this reason, the process of producing synthetic liquid
fuel from coal must now be viewed as the marginal one for the world
crude oil market.

Three large countries - the USA, the USSR and China - possess tremendous reserves of coal. These reserves will be sufficient for many centuries to come even if the existent energy consumption rate is maintained. The current costs of producing synthetic fuel from coal are greater than those for obtaining it out of oil extracted from deposits in the Persian Gulf area by an order but are nevertheless quite comparable to the existent level of oil prices on the world market.

The necessary condition for the production of gasoline from coal to take on the role of a marginal activity for the oil market is that this production can supply a sufficiently large share of the market. The latter is hindered by two obstacles: the first one being the tremendous amount of capital needed before commercial production will run because a production of this type becomes efficient only on a large-scale basis, and the second being the considerable time lag from the initial investment to real production. Moreover, there is very little available experience in the commercial production of gasoline from coal.

Presently, the world energy resource market is in a transitional period between a state when prices were based on the extraction costs of marginal oil deposits and a state when they will be based on the production costs of synthetic gasoline obtained from coal. This transitional period will most probably be over somewhere early in the twenty-first century. When it is over, it seems reasonable to expect stabilized prices for energy resources, which fact can be explained by coal reserves being practically inexhaustable.

During the transitional period, energy resource prices are strongly influenced by political factors. Exporting countries strive for maximum profits to be gained from their energy resources, at the same time taking into account the consequences of resource depletion and the impact of world inflation on the foreign currency accumulations of industrialized capitalist countries. Many industrial countries initiate special energy programmes aimed at the independence of energy supplies from abroad with nuclear energy and coal being the core of such programmes. The development of these programmes will be significantly dependent on energy supply to the world energy market and on the level of world oil prices. On the other hand, oil exporting countries, while forming their policy of supplies and prices, will undoubtedly take into consideration the extent of energy programme implementation in capitalist industrial countries. Besides this, rather perceptible energy resource deficits in capitalist countries are mostly probable during the transitional period, resulting from poor utilization of production capacities.

In our case, when we replace the aggregate resource by the energy resource, and the regional model by the RIM, it is logically more convenient to use the scheme of a jumping transition from one level of production costs to another instead of that of a gradual rise in resource prices with depletion of richer and more convenient deposits. It should be noted that in this case the most important role is played by the energy resource deficit expressed in physical units of the resource. This is that scheme which built the framework of the RIM presented in this paper.

## General Structure of the Model

Let us briefly review the model's structure. The RIM consists of two submodels. In the first one the influence of the energy resource deficit over the regional output, over the policy of investments into development of alternative energy resources and over the reduction of the GRP's unit energy consumption (energy saving) is described. (The modelling of energy deficit impacts on GRP is here principally similar to the "Fugi-ESCAP" model developed by A. Onishi and Y. Kaya of Japan /7/). The impact of the deficit over the regional output is determined by the coefficient specifying the share of the GRP actually produced in the potentially attainable GRP obtained from the corresponding production function /4, 5/. The deficit in its turn determines the amount of investments into development of alternative resources and into energy saving.

The demand of the energy resource in region "i" is determined by the GRP unit energy consumption coefficient. Supply is equal to the regional energy resource production and net imports of the resource. Production of the resource is dependent on investment into the development of alternative energy resources (according to the long-term character of the model). The regional import is calculated as a certain share in the world export of the energy resource. The latter is the sum of the energy resource exports over all the exporting regions with individual energy export equal to the difference between the extracted regional resource and the domestic demand for this resource with proper attention paid to policy factors. The proven reserves of the energy resource annually increase by a certain amount - the increase of the energy resource reserves for region "i" which is dependent on the energy resource world price with the latter dependent on the world energy resource deficit.

The second submodel of the RIM is actually a one-sector market model where for all regions (with the exception of oil-producing ones) the export price index is exogenously scenario-specified (and may be related to the inflation rate). This approach is an obvious simplification of the real process and is justified by methodological difficulties encountered when one tries to project the export price index far into the future as well as by the desire to build a model of reasonably small dimension. The import price index is calculated as the weighed average of export price indices with the weight coefficients being equal to the corresponding shares of exporting regions in the import of the importing region. Due to a lack of data, the import of goods and services in current prices is obtained as a regression estimate on the commodity import in current prices and the time variable. The regional commodity import is dependent on the GRP, export and total amount of aid received by the region. The regional commodity export is obtained as the sum of imports from this region into every other region. For the oil-producing regions the export price index is assumed to be dependent on the current energy resource world price. The rest of the regional export (the "invisible export") is determined in fixed shares of the total invisible world import. The resulting regional trade balance is a kind of indicator of "health" for this region's foreign trade activities.

Thus, the RIM presented here is a comprehensive econometric model where the concept of energy resource deficit is of major importance together with its overall influence over regional development. This influence is expressed in direct cutdowns of the potentially attainable GRP. It also influences the world price for the energy resource prices, resource extraction policies as well as energy programmes concerned with the development of alternative energy resources (the latter being a significant burden to regional economies). Taking into account the fact that the RIM is to cover a long-term time horizon, we have introduced a lot of dummy variables. These variables are used in the description of various policy factors as well as various structural changes which indispensably occur when the system modelled is in a transitional period.

The RIM presented below includes the principal indicators of regional foreign trade activities, of GRP production and utilization as well as those describing essential impacts of energy resource deficit over regional development.

## List of Variables

Let us introduce the following notation. A dash "-" above a variable means that this variable is exogenous. A wave "~" denotes a variable whose value is an input from a regional macro-model. A "˅" denotes that the variable is measured in current prices.

1. $V^i$ - actual Gross Regional Product (actual GRP).
2. $\tilde{V}^{i*}$ - potential GRP.
3. $C^i$ - regional consumption expenditure.
4. $\tilde{I}_k^i$ - regional fixed capital investments into goods and services production (excluding those into scientific development, education as well as governmental expenditures and national energy programmes).
5. $\bar{I}_N^i$ - investments into scientific development and education.
6. $\tilde{K}_M^i$ - depreciation.
7. $I_e^i$ - investments into the development of alternative energy sources.
8. $I_s^i$ - investments to bring about reduction of the GRP unit energy consumption (energy saving).
9. $\bar{I}_{TE}^i$ - total investments into development of alternative energy sources and energy saving.
10. $\tilde{J}^i$ - inventory investments
11. $\bar{G}^i$ - governmental expenditures (including fixed capital and inventory investments but excluding those into scientific development, education, energy programmes).
12. $\overline{DA}_i$ - official development aid to the developing countries.
13. $\overline{DF}_i$ - foreign private investments in the developing countries.
14. $E^i$ - export of goods and services.
15. $M^i$ - import of goods and services.
16. $e^i$ - energy resource deficit.
17. $D_e^{\ i}$ - regional demand for the energy resource.
18. $S_e^{\ i}$ - regional supply of the energy resource.
19. $e^i$ - regional GRP unit energy consumption coefficient.
20. $R_p^i$ - regional extraction of the energy resources.
21. $M_e^{\ i}$ - regional energy import.
22. $\tilde{U}^i$ - the regional technological level.
23. $M^{ij}$ - import from region "j" to region "i".
24. $Se^w$ - total world import (or total world supply for export) of the energy resource.

25. $P_e^w$ - price of the energy resource on the world market.

26. $\bar{z}_1^i$ - (dummy) variable reflecting changes in regional import policies.

27. $D^\Sigma$ - world energy resource deficit.

28. $\bar{z}_2^w$ - dummy variable reflecting changes in the world energy resource price formation mechanism. It is dependent on various non-economic factors.

29. $S_{ee}^i$ - export of the energy resource by region "i".

30. $R_{e_i}^i$ - proven reserves of the energy resource in region "i".

31. $R_e^i$ - increase of the energy resource reserves in region "i".

32. $M^i$ - import of goods and se rvices (on national account basis).

33. $M_c^i$ - commodity import of region "i" (on trade statistics basis).

34. $pM^i$ - regional import price index.

35. $\check{M}_i$ - import of goods and services in current prices.

36. $pE^j$ - export price index of exporter "j".

37. $(m^*)^{ij}$ - share of exporter "j" in the import of importer "i".

38. $m^{ij}$ - standardized share of exporter "j" in the import of importer "i".

39. $M^{ij}$ - commodity import by region "i" from region exporter "j".

40. $Ec^j$ - commodity export from region "j".

41. $\check{S}B^i$ - trade balance of region "i".

42. $\check{E}_{ot}^j$ - other export (invisible) of region "j" in current prices.

43. $\check{M}_{ot}$ - other import (invisible) of region "i" at current prices.

44. $\check{M}$ - world invisible import in current prices.

45. $\bar{z}_3^i$ - dummy variable reflecting regional trade policies.

46. $\bar{z}_4^{ij}$ - dummy variable reflecting changes in the shares of region exporter "j" in the commodity import of region "i" due to non-economic factors.

47. $\upsilon^i$ - actual GRP to potential ratio.

48. $\bar{\beta}$ - share of the investments into energy saving in the total of energy investments.

49. $(\varepsilon^*)^i$ - share of region "i" in the world import of the energy resource.

50. $\varepsilon^i$ - standardized share of region "i" in the world import of the energy resource.

51. $\bar{\omega}^i$ - annual rate of resource extraction

52. $z_5^i$ - dummy variable reflecting the influence of non-economic factors over the commodity import of region "i".

53. $e^j$ - share of region "j" in the invisible export.

54. $\check{E}_{ot}$ - world total invisible export at current prices.

## Model Equations (the subscript denotes time lag)

1. The accounting balance identity

$$v^i = c^i + \tilde{I}_k^i + \tilde{I}_N^i + \tilde{K}_\omega^i + I_{TE}^i + \tilde{J}^i + \tilde{G}^i + E^i - M^i$$

2. The relationship between the potential and actual GRPs.

$$v^i = \upsilon^i \tilde{v}^{ki}.$$

3. Share $\upsilon^i$ is a function on energy resource deficit $\Delta e^i$

$$\upsilon^i = f\ (1 - \Delta \frac{e^i}{De^i}).$$

4. The definitional identity for energy deficit

$$\Delta e^i = D_e^i - S_e^i.$$

5. $D_e^i = e^i \tilde{v}^{*i}$

6. $S_e^i = R_r^i + M_e^i.$

7. It is assumed that the GRP unit energy consumption coefficient is dependent on the level of technological advancement as well as on investment in energy saving and the growth rate of the regional economy.

$$e^i = 4^i\ (\tilde{U}^i,\ I_s^i,\ \Delta v^i / v^i).$$

8. $I_e^i = \bar{\beta} \bar{I}_{TE}^i.$

9. $I_s^i = (1 - \bar{\beta})\ \bar{I}_{TE}^i.$

10. $\bar{I}_{TE}$ - total investments into the development of alternative energy resources and energy saving are scenario-specified.

11. $R_r^i = \phi^i \, (I_e^i)$.

It is assumed that for the transitional period domestic energy resource extraction in regions-importers will mainly be determined by investments into the development of alternative energy resources (i.e., by the rate of energy programmes implementation).

12. $M_e^i = \varepsilon^i S_e^\omega$.

13. $\varepsilon^i = \varepsilon^{*i} / \sum\limits_{i} \varepsilon^{*i}$.

14. $\varepsilon^{*i} = a_{0.14}^i + a_{1.14}^i \, V_{-1}^i + a_{2.14}^i \, r_e^\omega + a_{3.14}^i \, \bar{t} + a_{4.14}^i \, \bar{Z}_1^i$.

We assume that the share of region "i" in the world import is dependent on the GRP produced in the preceeding year as well as on the world energy resource price, the time variable and various non-economic factors which are introduced by the dummy variable $\bar{Z}_1^i$.

15. $r_e^\omega = a_{0.15}^i + a_{1.15}^i \, D^\Sigma + a_{2.15}^i \, \bar{t} + \bar{a}_{3.15} \, \bar{Z}_2^\omega$.

The world energy resource price is a function of the world energy resource deficit, the time variable and various policy factors.

16. $D^\Sigma = \sum\limits_{i \in I} e^i$.

I denotes the set of all regions importing energy.

17. $S_e^\omega = \sum\limits_{i=J} S_{ee}^i$.

J - a set of all regions exporting energy.

18. $S_{ee}^i = a_{0.18}^i + a_{1.18}^i \, [\bar{r}R_{e_{-1}}^i - D_e^i] + a_{2.18}^i \, \bar{Z}_3^i$.

The energy resource export of region "i" is a function potential export $[\bar{r}R_{e_{-1}}^i - D_e^i]$ and some policy factors.

19. $R_e^i = R_{e_{-1}}^i - R_{r_{-1}}^i + \Delta R_e^i$.

20. $\Delta R_e^i = a_{0.20}^i + a_{1.21}^i \, r_{e_{-1}}^\omega$.

The increase in energy reserves in region "i" is mainly dependent on the world energy resource market price of the preceding year.

21. For energy exporting regions we assume that $\overset{*}{V} = V$.

22. For the oil-producing Middle East, GRP is scenario-specified.

23. $\check{M}^i = a^i_{0.23} + a^i_{1.23} M^i_e + a^i_{2.23} \bar{t}.$

24. $\check{M}^i_e = M^i_e \, rM^i.$

25. $rM^i = \sum_{j \in J} m^{ij} \, rE^j.$

26. $SB^i = \check{E}^i - \check{M}^i.$

27. $m^{ij} = (m^*)^{ij} / \sum_{j \in J} (m^*)^{ij}.$

$J_i$ - denotes the set of all regions exporting into region "i".

28. $(m^*)^{ij} = a^i_{0.28} + a^i_{1.28} V^i_{-1} a^i_{2.28} \tilde{U}^i_{-1} + a^i_{3.28} \bar{t} + a^i_{4.28} \bar{z}^{ij}.$

The share of exporter "j" in the import of importer "i" – $(m^*)^{ij}$ – is dependent on the regional export capacity, the time variable and various policy factors taken into account by means of the dummy $\bar{z}^{ij}_4$.

29. $M^i_c = ki \, (V^i) \, \beta i \, [E^i + (DA+DF)^i]^{ri}_{-1} \, {}^{-S_i}_{z_5}.$

30. $M^{ij} = mij \, M^i_c.$

31. $E^j_c = \sum_{i=Ij} M^{ij}_c.$

Ij denotes the set of all regions importing the energy resource from the j-th region-exporter.

32. $\check{E}^i_c = E^i_c \, rE^i.$

The export price index for every region, excluding that of the oil-producing regions, is scenario-specified.

33. The export price index for the oil-producing region is determined by the energy resource world price level.

$$rE^i = a^i_{0.33} + a^i_{1.33} \, r^\omega_e.$$

34. $\check{E}^i = \check{E}^i_c + \check{E}^i_{ot}.$

35. $\check{M}^i_{ot} = \check{M}_i - \check{M}^i_c.$

36. $\sum_i \check{M}_{ot} = \check{M}_{ot}.$

37. $\check{M}_{ot} = \check{E}_{ot}.$

38. $\check{E}^i_{ot} = 1_i \check{E}.$

39. $E^i = \check{E}^i/rE^i.$

40. $M^i = \check{M}^i/rM^i.$

References

1. D.M. Gvišiani. Globalnoje modelirovanie: kompleksnij analiz mirovogo razvitija. "Problemi mira i socijalizma", No. 8, avgust 1978.

2. D.M. Gvišiani. Metodologičeskie problemi modelirovanija globalnogo razvitija. Moskva, 1977.

3. S.V. Dubovskij. Sistema modelej processov globalnogo razvitija.

4. S.V. Dubovskij. Proizvodstvennij funkcional s endogeniji i upravljaemiji naučno-tehničeskim progressom. Sb. trudov vniisi, No. 9, M. 1978.

5. V.A. Gelovani, S.V. Dubovskij, V.V. Jurčenko "Modelirovanie dolgosročnoga razvitija regiona". Dokladi Akademik Nauk SSSR, 1978 g.t. 238, No. 3, str. 534-537.

6. Jean-Marie Chevalier. Le Nouvel Enjeu Pétrolier. Paris, 1973.

7. A. Onishi. Impacts of Raising the Oil Price on the Future of the ESCAP. Member Countries. Toyko, November 1979.

# CONTRIBUTION TO THE SIMULATION MODELLING
## OF AN ECONOMIC SYSTEM

Miloš Rajkov, Slobodan Andrić,
Zdravko Šišarica, Ljubiša Rajin

Faculty of Organizational Sciences,
Belgrade, Yugoslavia

In spite of the efforts which are invested in the formulation of the economic situation, we are too often witnesses to the ensuing consequences of proposed and adopted policies which are either unexpected or it may be that events protrude the scope anticipated. Having these facts in mind, both theory and practice are developing for the sake of eliminating deviations from expected results.

It is undoubtedly accepted that the economic system is complex, non-linear and of a higher order. Analysis of the said system and its management is today taking priority in investigations. Society growth is less and less given out to chance and ungoverned processes. Exponential growth, which is characteristic for economic systems with small starting deviations, afterwards results in significant differences between objectively possible and exhibited results.

A qualitative approach in the analysis of social reproduction has its basis in the Marx theory of reproduction /1/.

This theory most frequently starts out from the standpoint of conditions for the normal and complete exchange of the manufactured good, from the standpoint of the establishment of balance and convenient proportion between total supply and demand. Theoretical analysis of reproduction presupposes an already-defined multitude of economic categories.

Real reproduction executed in the actual economic system deviates from the theoretical one. The harmony between total supply and demand is not realized, exchanges are fewer than those which might have been expected from labour invested presently and in the past. The question is posed as to whether it is possible to complete theoretical analyses with new ones, e.g., with the development of a mathematical model of the economic system serving the purpose of the choice of the social strategy of the more harmonious development of supply and demand.

The beginning of the quantitative approach to the research of reproduction problems could even be traced in Marx´s works on reproduction schemes. This basic approach can be developed today by utilizing the latest scientific and theoretical knowledge in the field of cybernetics and systems theory. A real economic system can be easily transposed into a simulation mathematical model and reproduction can be analyzed from its results, not just theoretically but actually. On the basis of such a model it is feasible to weigh the influence of all economic policy subjects which might have a bearing on results anticipated in the future.

## The Economic System: Fundamental Thought on the System and Its Aims

The economic system would be viewed through the flow of the consumption of goods, means for work, the object for work, funds and information. All the system´s elements are found on the corresponding flows and belong to the subsystems which make up the economic system (Chapters I.a, I.b and II, according to Marx).

System structure analysis leads us to the conclusion that the system is made up of a combination of feedback loops. The behaviour of such a system is dominated by the relationship between its elements (policy and decision criteria) and latency in the propagation of the change impulse.

The economic system is a set of inputs, outputs, elements and their relationships. Relations connect the entire system into a whole. Inputs and outputs are various and are carriers of material, money, energy and information. Inputs and outputs determine the connection with the environment which may be the natural system (natural ore and energy wealth) or other organizational systems (economic systems with which exchange is executed, socio-political systems, etc.). Some of the input-output elements can significantly influence system behaviour in that they transmit impulses born in conjunct systems. For the system whose behaviour depends solely on its dynamics we say that it is a closed system. It is possible to investigate it and reveal the greatest number of courses for its changes in behaviour.

In spite of the fact that it is impossible to find the closed system among organizational ones, for the economic system of wider social area we might say that by its degree of closure it is far more closed than any other organizational system. Therefore for such a system we say that it is under the influence of input, output and its own dynamics. The system reacts to all three dynamics in conjunction with its inherent nature, i.e., with its own structure /2/.

The development of economic relations in a society has led to a clear definition of the economic system's raison d'etre. It can be defined most briefly as a system with the task of satisfying the needs of a society (common and personal ones) with production potentials. It is necessary to take into consideration all disposable natural resources, labour which is invested in production and man's capacity to invent and develop. Necessities are permanently increasing and changing their structure. These changes have to be followed by changes in production potentials. The economic resources necessary for enlarged demands and production arise from goods turnover, which is executed according to known economic laws and social principles.

The economic system has been constituted, through evolution, in a manner which assures its survival, growth and development. Social relations' evolution, development of knowledge and human capabilities, enlarged demand and production possibilities condition the ever-lasting change in the economic system.

Adaptation to new conditions is one of the system's dynamic resources. Impulses are transmitted to the whole system and trigger the changes. They may be positive or negative in relation to the satisfaction of social needs. One of the most important questions is how to measure the successfulness the economic system. In spite of the fact that it arose from the need to harmonize distribution of the newly created work value with the consumption and growth of production potentials, this system's functioning mechanism is not such as to always expect satisfactory results from the harmonization. Changes in the system arise in a twofold manner. Problems seek change in the behaviour of individuals and groups. It is exercised in an unorganized, unbridled manner. The desire to improve the situation by changes can be the source of favourable and unfavourable changes in a system. The other change is realized by designing the system's changes with advance laboratory investigation of its qualities.

Systems can be changed, and they should be such as to have:
- the best utilization of available resources,
- the easiest harmonization (quantitative and timely) for satisfying the various needs.

In order to manage the economic system for the sake of establishing a balance between various needs and possibilities, it is necessary to have developed reasoning power. The search for *one goal* is indeed a search for a non-existent formula to replace it. The system's nature demands a multitude of goals. During their formation, the number of alternatives should be decreased by the adoption of one clear standpoint and by the establishment of a firm data base.

It is anticipated in these conditions that the satisfaction of each singular need demands the least investments. This is a clear task for a design. However, when we start with the formulation of a new one, the question remains of how to judge a new solution in relation to other possibilities. Let us examine, therefore, the general economic goals which could be defined as:

- a growth of social efficiency over the total invested labour fund,

- a growth of necessity efficiency of all material resources,

- faster growth of the areas to be the main bearer of the new techniques,

- the maintaining of dynamic proportions between economic areas and regions.

Thus, formulation goals are clear but insufficiently precise for decision making.

Determination of the number of goals, which ranges between one and many, has been given to the reasoning power which, regardless of its magnitude, may not always be able to select the right one but be able to treat them accordingly in a list.

It is not a small number of cases where the goals and their embodiment have brought about the phenomenon of new problems and stagnation. This happens because the goals are not clearly defined and their influence on results can trigger various impulses and adjoints in the conjunct feedback circuits. All events in the realization of the goals are happening in a system of connected feedback circuits. Knowledge of the structure of a system and utilization of the simulation can contribute to determining that combination of goals which, in the slim limits of permissible results´ variations, can provide optimum realization. Actually, a model could clear out a phenomenon of system function and all causes originating from realization of the goals:

- the system´s behaviour can be forecast in a laboratory,

- the advantage of the decisions made can be measured,

- the quality of the managers´ economic reasoning can be analyzed by itself.

The economic system analysis issues the conclusion that it is not the aim of these systems to transform inputs into the desired outputs. It is a system which should provide an internal balance and relationships between elements and subsystems. Inputs, outputs and their magnitude are the reflection of a conscious desire to harmonize needs and possibilities as a part of a system in a manner which is the most appropriate for attaining goals in the shortest time. From the

above-cited, it is clear that the goals shall relate to the system
states to be achieved. Namely, capacity, consumption, inventory, manpower and other states should appear in numbers, individually or in a
combination of different states. The quantity of magnitude of some
elements, which appears as one from the whole set of goals, is defined
on the basis of real status and supposition which appear from analyzing
the environmental factors. It has been pointed out that the possibility of goal realization can be evaluated in the laboratory even before
these goals are accepted and measures programmed for their realization
brought into harmony with expected possibilities.

## The Economic System´s Structure - System Flow Diagram

If the interpretation of the economic system as a conjunction
of the feedback loops is accepted, we can complete the explanation of
Marx´s scheme of the process of capital´s circular motion: $N-R_{rs}^{sp}$ ...
P ... R ... N. The scheme contains within itself money, material and
goods which constitute different paths in a system. Each of the paths
forms, with the corresponding information flows, the conjunct feedback
purpose. Marx deliberated on liasons between the feedback circuits.
Money is not transformed into material but, with its flow, influences
the formation of materials flow. An equivalent part of money passes
from one chapter to another. These movements of material and their
transformation from the lower to the higher utilization value is followed by the simultaneous movement of money from one region of consumption to trade and production. This difference in interpretation can
be easily identified by relating charts 1. and 2. Of the first, it
might be said that the process of the circular motion of capital is
accurately represented together with Marx´s transformation. In chart 2,
where the system structure is developed in more detail, interpretation
of the said circular movement of capital should be more fully comple-
mented. This supplement shows that the starting capital N consists of
a greater number of independent variables on different money paths. If
we say that the starting capital is the sum of capitals for the work
means and production means, then we can follow each partial capital on
its path. The first part of capital from other departments flows towards the first a(Ia), and the second part from other departments flows
towards the first b (Ib), and labour costs from all departments flow
into the purchasing power. Accordingly, the first part is exchanged
at its end as the basic means, as goods, the second one for material
as goods and the third as personal consumption goods. This supplement-
ary interpretation does not alter Marx´s scheme of the circular motion

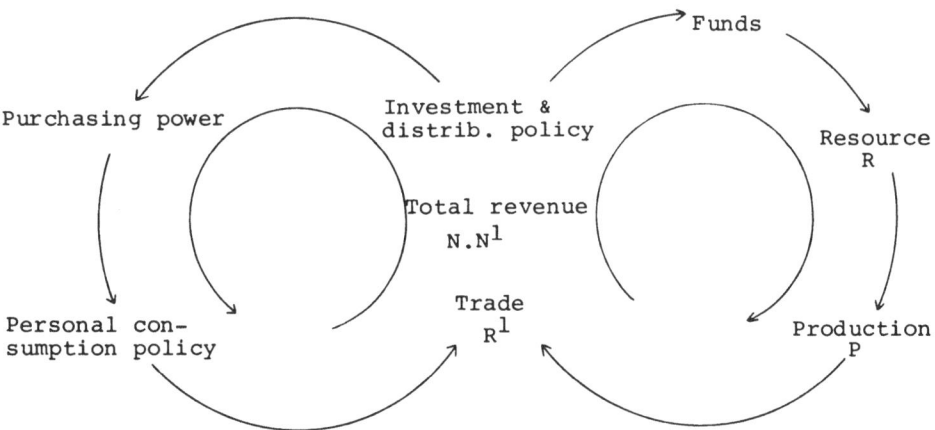

Funds

Purchasing power   Investment &
                   distrib. policy          Resource
                                            R

                   Total revenue
                   N.N$^1$

                   Trade
Personal con-      R$^1$          Production
sumption policy                  P

Chart 1. Circular flow of capital represented in a scheme of the
economic system

Invest. funds la

Income distrib.        Invest.
policy la              policy la

Income la                              Production means
                                       (import)

Total revenues    Capacity
N la              Re la
                                       Work object
                                       Rm la

                                       Production la

Purchasing power  Trade of the
                  production
                  means la
                                       Manpower
                                       R1 la
            Invest. policy

            Invest funds    Capacity
Consumption
policy      Inc. dist.
            policy II                  Manpower R1 II

Savings     Income
                                       Production II
            Tot. rev.
            N II     Personal          Work object
                     consump. R II     Rm II

                     Goods

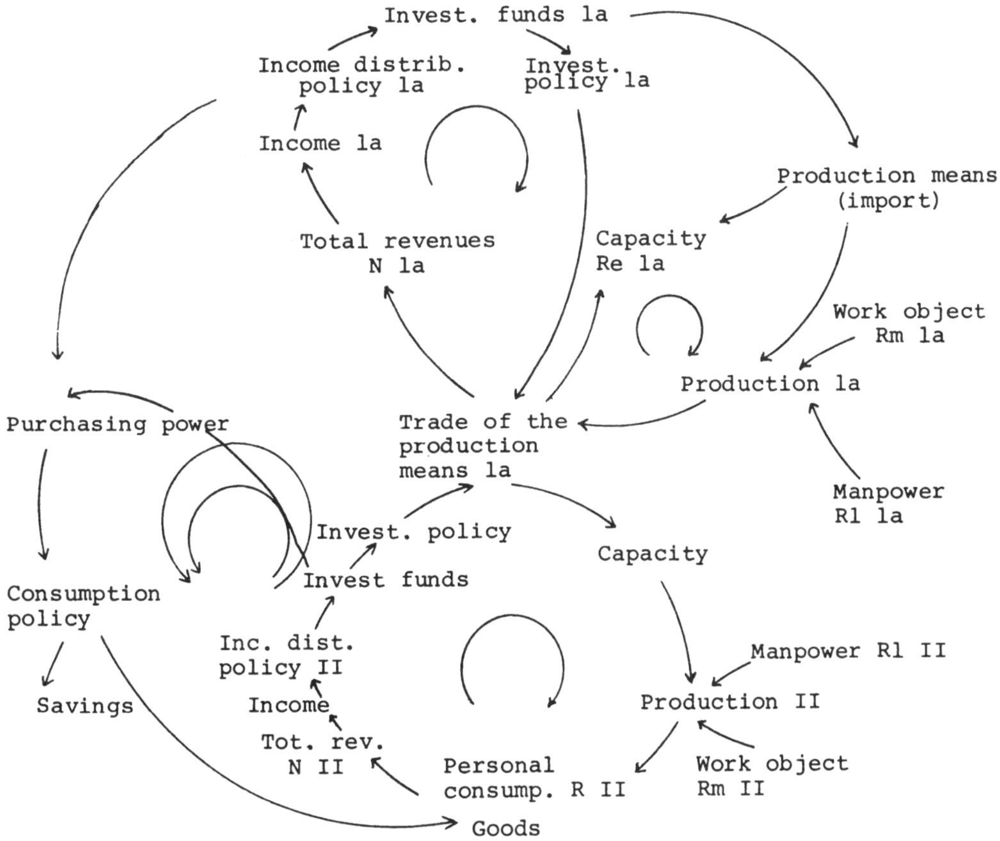

Chart 2. Circular flow of capital represented in a developed scheme of
the economic system

of capital but brings to it a new dimension harmonized with the latest revelations acquired through application of the system theory.

Marx´s schemes of introduction are the closer concretization of the circular motion process scheme of capital for each chapter of the individual economic system.

A more detailed structure of the economic system depicted in chart 1 is presented in chart 3 /3/. The following flows in the system are now clearly observed:

- money flow at population,
- money flow for investment in fixed assets,
- capacities flow of entire economy,
- commodity flow,
- information flows which bind them in totality.

In each money flow at population, there is a state variable pointing to the amount of money possessed by the population (MP) which together with input money from sales (IPP) represents the population purchasing power (PP). The spending of money depends on commodity sales activity. In this case, the other forms of money output are not supposed. Population purchasing power makes impacts upon the volume of commodity sales which simultaneously depends on the potential sale of commodity goods (PSL). The volume of commodity stock determines the amount of salable goods. Therefore, money will be spent totally by the population providing that there are enough commodities, or as much as there are goods available. The rest of the money is accumula-ted as money at population. Gross national income (I) is created from the magnitude of sales and efficiency of production technology. The amount of total income influences the magnitude of money input at pop-ulation. The hitherto noted elements form positive feedback loop A on chart 3.

The relationship between potential sales and purchasing power influences the investment policy of society for expansion of production capacities. Greater demand over supply results in enhanced investment for new production capacities. The magnitude of input and rate of accumulation for gross investment fund (DRI) determine the input of money (IGIF) into gross investment fund (GIF). From this fund the money output depends on new capacities put in exploitation (FCC). The investment policy regulates the way in which the amount of money in the investment fund will influence the outset of new capacities´ con-struction (SC). This parameter regulates the likely volume of con-struction commencement (CSF). The commenced construction of capacities causes input of new capacities in exploitation (FCC) after a time lag related to the duration of capacity construction (TCC). New capacities

225

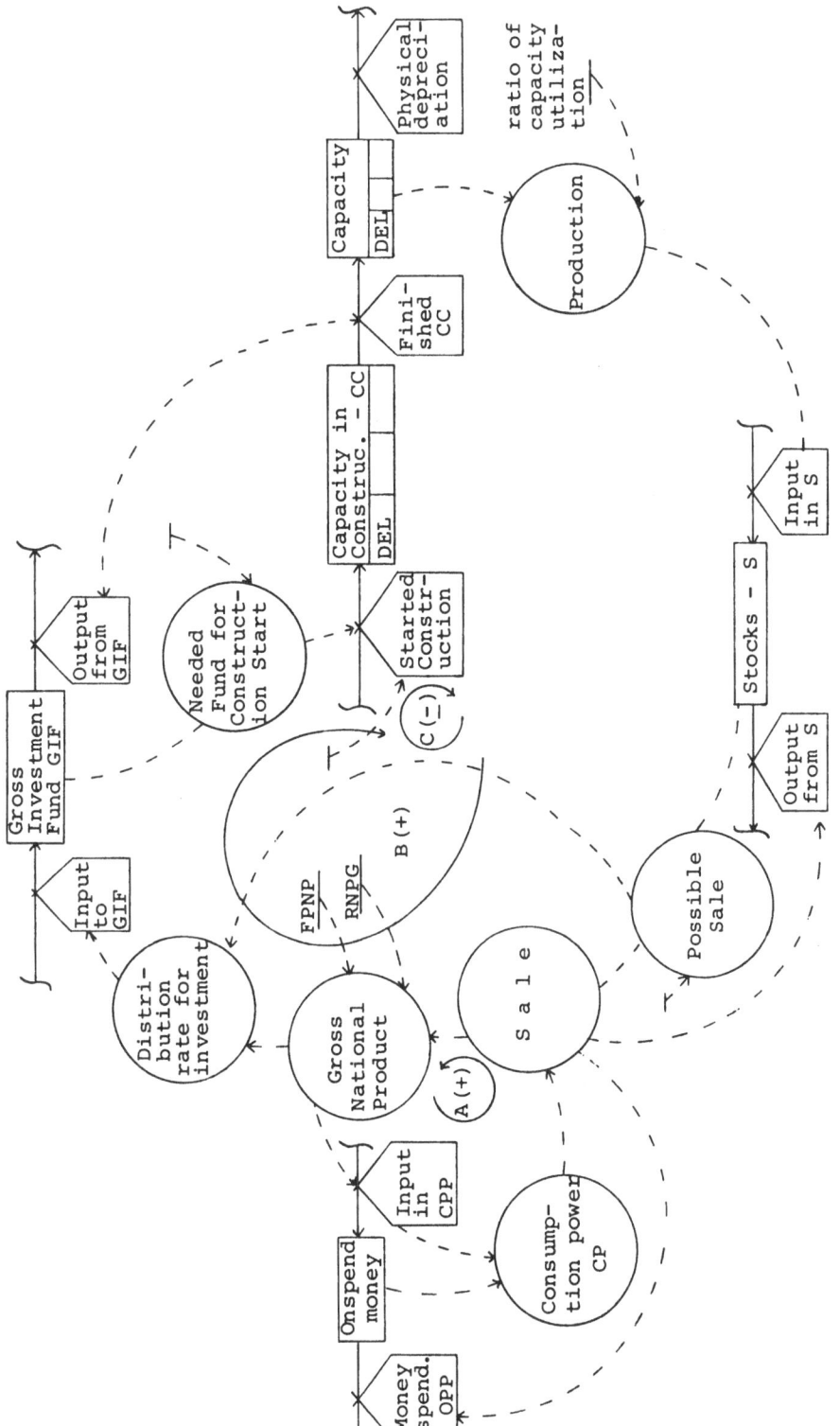

Chart 3. Flow diagram of economic system

increase those in exploitation (C). Available capacities diminish due
to physical depreciation (FD) which is subject to the magnitude of
available capacities and average time of those in exploitation (TFD).
The volume of available capacities predetermines production (P) which
changes commodity stocks for sale (S). Stocks diminish depending upon
the volume of goods sold (SO).

The cited elements of the system form two conjuncted feedback
loops (B and C on chart 3). Loop B is positive and supports system
growth, while loop C is negative and controls increment in loops A and
B, maintaining a balance between supply and demand.

The structural model depicted is the ultimate simplification
of the real structure of the economic system. Tracing the logic presented in respect of the formation of this structure it is easy to come
across the structure showing Marx´s departments separately as well as
the branches and activities of the concrete economic system.

## Mathematical Model of the Economic System

The saying that theoretical and abstract considerations are
apart from application in economic practice, that they are something
unconnected to economic policy and useless for economic policy, should
be discarded as erroneous. Such mistakes, when speaking of mathematical sciences, have been developed most particularly in recent years.

Theoretical analysis of the observed practical problems of
the economic system is posed as an imperative to contemporary researchers whose task in regard to the construction and management of the
economy differs from those in Marx´s era, although we find the best
approach to the investigation of economic systems in his assumptions.

In a new approach to the economic system, the starting point
should be the latest knowledge of science on the system´s management.
These revelations could discover unknown factors to surprise managers
and govern complex and composite problems of data processing and utilization of adequate mathematical models. This new approach indicates
the system´s problems, its structures, parameters, interrelations of
elements and its partial or integral behaviour.

In model formulation, we should start from the following
economic system assumptions:

- the system is a composite organization and functioning of
various states and flows controlled by the feedback loops,

- the system is goal oriented,

- functioning is consequenced by various incorporated management and executive activities.

The system structure arises in the conjunction of feedback circuits from the flows of money, material, equipment and information. Information flows comprise policy and criteria for decision making, which influence changes in the state of all elements. The interdependence of system elements forms feedback loops. The behavioural change impulse arising in any part of a system is transmitted to the whole system. Behavioural change can be significant and its duration should be ascertained.

The mathematical model could correspond to the system structure and could imitate activities in a real world. Using this model, we can quantitatively follow events effected by management actions and by the nature of the very system. In this manner feasible realizations can be assessed - reaching out the goals of the economic system. Another benefit of the mathematical model is the possibility of evaluating a stability of system to the various impulses from the environment or from the system. The economic system´s behaviour is interpreted in various theoretical ways. The greatest variety can be revealed in the interpretation of the economic cycle dynamics. It should be admitted that this is a most peculiar interpretation of a cycle and the closest to a true answer to the questions: what is the cycle´s cause and why is it such? This interpretation says that "as the wooden horse swings with a frequency and amplitude which partly depend on its internal nature (size, weight), so in such a manner shall the economic system react to the fluctuation of external factors in accordance with its internal nature. Both external and internal are important for the explanation of the cycle".

The researcher´s basic task consists exactly of revealing the internal nature of the economic system. Contemporary knowledge on its structure clearly indicates that such a system behaves in various ways under the influence of internal or external changes. The economic system´s constitution can be such that if it is once brought out from a state of equilibrium it cannot reach new equilibria. Due to the economic system´s composite structure, its behavioural feature fluctuates, is probable and unstable.

For non-linear, higher order systems with activities taking place in discrete periods, the mathematical simulation model is the most suitable tool for the investigation of system dynamics and results realizable under the influence of management decisions. As the economic system has a structure made up of feedback circuits and in order to represent such a structure, the mathematical model is actually a system of differential linear equations.

The simulation model of the economic system has some advantages over other mathematical models developed up to now.  Such a simulation model represents the real nature of the economic system and it can be utilized for the long-term tracking of system activities under the influence of a given economic policy and decision criteria.

The dynamic system simulation is an outstanding instrument for efficient exploration, projection and conduct with large and complex systems as it is with the economic system.  For mathematical purposes we employed the system of first order difference equations.  The numerical method of simulation to solve this system of equations can be either Euler´s or Kütte-Merson´s method /4/.

In this paper we shall present only a part of the complete simulation model.  This is formulated by utilizing the already-described flow diagram as well as by ascertaining the mathematical form of relationship among system variables.

This part of the model relates to the formulation of the rate of singling out for funds and variables on money flow for investment in the following way:

```
        GO TO (40,50), INT (PR)
        P12 = 10
        E 20.K = TABLE (T1,E1.K/E6.K,.4,1.6,.1)
40
        E20.0 =.2
        T1/.07,.11,.14,....
50
        E20.K = Table (T2,E1.K/E6.K,.4,1.6,.1)
        T2/.07,.08,.09,....
```

and describes two likely policies in singling out for the construction of new capacities.  One is such that for small changes in the demand/supply relationship the rate of means for investment shifts slightly while the second is just the opposite.  The programme language used here has the characteristic feature that instruction for analysis of the model with all other unchanged suppositions simultaneously provides the system behaviour if the first or second policy is changed.  In this way the analysis of system behaviour is ultimately quick and conforms.

The model continuation is as follows:

```
60      E19.KL = E20.K E4.K
        E17.K = E17.J +DT (E19.JK-E18.JK)
        E17.0 =600
        E16.K =E17.K P8
        P8=.3K
```

and describes the amount of money in the investment fund (E17) as well as the volume of means required for capacity construction (E16).  The model with a description of the significance of variables and parameters of the system is submitted in Supplement 1.

## Model Exploration of System Behaviour

The economic system is described with the model which comprises 10 state variables, 15 rate variables and 10 parameters. Explorations of behaviour are first carried out in respect of model estimation and its capability to present the behaviour of the economic system. This job, with the application of SDS programme language, has been processed very efficiently and quickly. Out of a huge number of explorations, the outcome of changes of parameter values and starting values of state variables, we selected three specific scenarios which in themselves, providing the same system structure, give a different system behaviour. The scenarios are related to the change of three parameters and their value combinations:

### S C E N A R I O S

| Parameter | I | II | III |
|---|---|---|---|
| P2 - Rate of GNP Growth | 1. | .95 | .95 |
| P7 - Policy of GIF Use | .34 | .34 | .30 |
| P8 - Policy of CS versus NFCS | 1. | 1. | 3. |

The corresponding behaviours are presented in chart 4. We succeeded, with the same structure, in obtaining the characteristic behaviour of economic systems. The line marked I relates to the first scenario and shows economy with GNP growth. Line II relates to the second scenario and simulates economy with no growth but stable. Line III relates to the third scenario and simulates economy with no growth and no stability.

Explorations have been resumed with very interesting conclusions. For example, the two possible policies of singling out to investment funds depicted and described influence a different growth of GNP (chart 4). It is evident that a quick reaction in change of demand/supply relations in directing means into investment supports a faster growth of the economic system. Parameters and tables in the model are the reflection of economic policy. The outputs clearly show that different policies in the same system impact on various system behaviours and effects of performance. The model clearly demonstrated that it could be of use in the evaluation and choice of economic policy measures which, in the long run, have to provide better performance results.

   The mathematical model presented is not reliable for a concrete economy although the relations of starting values of state variables are real.  The relations´ inter-elements are close to reality,
although they can be formulated differently in a concrete system.

   The results obtained encourage further investigations and
building up of the mathematical model to be less aggregated and, accordingly, showing a more real system behaviour.

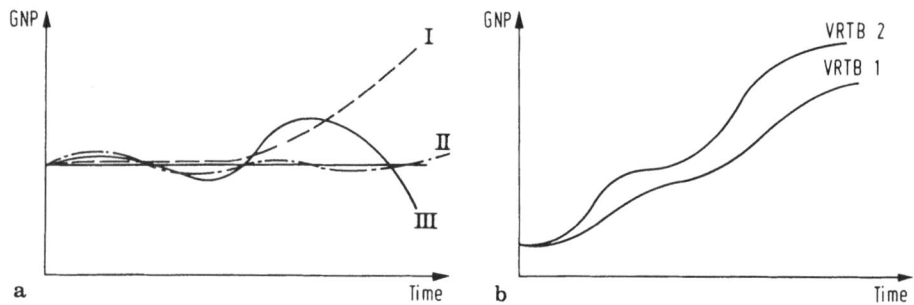

Chart 4. Possible forms of economy behaviour subject to the magnitude
     of parameters related to investment policy and production
     efficiency.

References

1. Karl Marx, Capital, First Book, Chapter 21.

2. Paul Samuelson, Economics: An Introductory Analysis, McGraw Hill,
   1967, New York.

3. Jay Forrester, Industrial Dynamics, MIT Press, Cambridge, Mass.
   1964.

4. M. Rajkov, S. Andrić, SDS - System Dynamics Simulation Interactive
   Language, 1980, Decus Europe, Proceedings.

```
MODEL MNE001
SEMINAR ON GLOBAL MODELING - DUBROVNIK 1980.
MODEL OF NATIONAL ECONOMY (VERZION 1)
E1.K     PP      A      DIN      PURCHASING POWER
E2.KL    IPP     B      DIN/GOD  INPUT IN PURCHASING POWER
E3.KL    OPP     C      DIN/GOD  OUTPUT FROM PURCHASING POWER
E4.K     GNP     D      DIN/GOD  GROS NATIONAL INCOME
E5       SL      E      DIN/GOD  SALE
E6.K     PSL     F      DIN/GOD  POTENTIAL SALE
E7.K     S       G      DIN      STOCKS
E8.KL    SI      H      DIN/GOD  INPUT IN STOCKS
E9.KL    SO      I      DIN/GOD  OUTPUT FROM STOCKS
E10.K    P       J      DIN/GOD  FINISHED PRODUCTION
E11.K    C       K      DIN      CAPACITIES
E12.KL   FD      L      DIN/GOD  PHYSICAL DEPRETIATION OF CAPACITIES
E13.KL   FCC     M      DIN/GOD  NEW CAPACITIES
E14.K    CC      N      DIN      CAPACITIES UNDER CONSTRUCTION
E15.KL   SC      O      DIN/GOD  STARTED CAPACITY CONSTRUC
E16.K    NFCS    P      DIN/GOD  MEANS NEEDED FOR CAPAC. CONSTRUCTION
E17.K    GIF     Q      DIN      GROSS INVESTMENT FUND
E18.KL   OGIF    R      DIN/GOD  OUTPUT FROM GIF
E19.KL   IGIF    S      DIN/GOD  INPUT IN GIF
E20.K    DRI     T      DIN/GOD  DISTRIBUTION RATE FOR GIF
E21.K    IIPP    U      DIN/GOD  UNIDENTIFIEND INPUT IN PURCH.POWER
P1       CVG     -      -        CONSTANT VOLUME OF GNP
P2       RGG     -      -        RATE OF GNP GROWTH
P3       DPS     -      -        DEPANDANCE OF POTENT.SALE FROM ST.
P4       RCP     -      -        RELATIONSHIP OF CAP.AND POT.PROD.
P5       TPD     -      -        TIME OF PHYS.DEPRETIATION OF CAP.
P6       CCT     -      -        CAPACITY CONSTRUCTION TIME
P7       VOM     -      -        VOL. OF CONSTR.OUTSET RESP AV.MEANS
P8       AMC     -      -        AV.CASH MEANS FOR CAP.OUT.CONSTR.
P9       IKC     -      -        INPUT KIND CHOICE IN SYSTEM
P10      MUS     -      -        MASS OF UNIDENTIFIELD SUIT.DEMAND
P11      ATU     -      -        APPEARANCE TIME OF UNID.SUIT.P.P.
P12      TKIN    -      -        TABLE KIND
T1       TBDR    -      -        TABLES OF VALUES OF DIST.RATE F.G.
 VOLUME OF STEP SIMULATION
DT=1.
 NUMBER STEPS OF SIMULATION
BS=75.
 NUMBER ELEMENTS OF MODEL
BE=21.
 NUMBER PARAMETERS OF MODEL
BP=12.
 MODEL
E1.K=E1.J+DT*(E2.JK-E3.JK)
E1.0=1000.
E7.K=E7.J+DT*(E8.JK-E9.JK)
E7.0=3000.
E6.K=E7.K/P3
E6.0=1000.
P3=3.
IF(E1.K-E6.K) 20,20,10
10       E5.K=E6.K
E5.0=1000.
GO TO 30
20       E5.K=E1.K
30       E4.K=P1+P2*E5.K
E4.0=1050.
P1=50.
P2=1.
E21.K=STEP(P10,P11)
```

```
        E21.0=0.
        P10=10.
        P11=5.
        E2.KL=E4.K+P9*E21.K
        P9=0.
        E3.KL=E5.K
        GO TO (40,50),INT(P12)
        P12=1.
40      E20.K=TABHL(T1,E1.K/E6.K,.4,1.6,.1)
        E20.0=.2
        T1/.07,.11,.14,.16,.18,.19,.20,.21,.22,.24,.26,.29,.33/
        GO TO 60
50      E20.K=TABHL(T2,E1.K/E6.K,.4,1.6,.1)
        T2/.07,.08,.09,.11,.14,.18,.20,.22,.26,.29,.31,.32,.33/
60      E19.KL=E20.K*E4.K
        E17.K=E17.J+DT*(E19.JK-E18.JK)
        E17.0=600.
        E16.K=E17.K*P8
        E16.0=180.
        P8=.34
        E15.KL=E16.K*P7
        P7=1.
        E14.K=E14.J+DT*(E15.JK-E13.JK)
        E14.0=600.
        E13.KL=DELAY(3.,E15.JK,P5)
        P5=3.
        E11.K=E11.J+DT*(E13.JK-E12.JK)
        E11.0=2000.
        E12.KL=DELAY(3.,E13.JK,P6)
        P6=10.
        E10.K=E11.K/P4
        E10.0=1000.
        P4=2.
        E8.KL=E10.K
        E9.KL=E5.K
        E18.KL=E13.JK
        CALL SDS
```

# Lecture Notes in Control and Information Sciences

Edited by A. V. Balakrishnan and M. Thoma

Vol. 22: Optimization Techniques
Proceedings of the 9th IFIP Conference on
Optimization Techniques,
Warsaw, September 4–8, 1979
Part 1
Edited by K. Iracki, K. Malanowski, S. Walukiewicz
XVI, 569 pages. 1980

Vol. 23: Optimization Techniques
Proceedings of the 9th IFIP Conference on
Optimization Techniques,
Warsaw, September 4-8, 1979
Part 2
Edited by K. Iracki, K. Malanowski, S. Walukiewicz
XV, 621 pages. 1980

Vol. 24: Methods and Applications
in Adaptive Control
Proceedings of an International Symposium
Bochum, 1980
Edited by H. Unbehauen
VI, 309 pages. 1980

Vol. 25: Stochastic Differential Systems –
Filtering and Control
Proceedings of the IFIP-WG7/1 Working Conference
Vilnius, Lithuania, USSR, Aug. 28 – Sept. 2, 1978
Edited by B. Grigelionis
X, 362 pages. 1980

Vol. 26: D. L. Iglehart, G. S. Shedler
Regenerative Simulation of Response
Times in Networks of Queues
XII, 204 pages. 1980

Vol. 27: D. H. Jacobson, D. H. Martin, M. Pachter, T. Geveci
Extensions of Linear-Quadratic Control Theory
XI, 288 pages. 1980

Vol. 28: Analysis and Optimization of Systems
Proceedings of the Fourth International
Conference on Analysis and Optimization of Systems
Versailles, December 16–19, 1980
Edited by A. Bensoussan and J. L. Lions
XIV, 999 pages. 1980

Vol. 29: M. Vidyasagar,
Input-Output Analysis of Large-Scale
Interconnected Systems –
Decomposition, Well-Posedness and Stability
VI, 221 pages. 1981

Vol. 30: Optimization and Optimal Control
Proceedings of a Conference Held at
Oberwolfach, March 16–22, 1980
Edited by A. Auslender, W. Oettli, and J. Stoer
VIII, 254 pages. 1981

Vol. 31: Berc Rustem
Projection Methods in Constrained
Optimisation and Applications
to Optimal Policy Decisions
XV, 315 pages. 1981

Vol. 32: Tsuyoshi Matsuo,
Realization Theory of
Continuous-Time Dynamical Systems
VI, 329 pages, 1981

Vol. 33: Peter Dransfield
Hydraulic Control Systems –
Design and Analysis of Their Dynamics
VII, 227 pages, 1981

Vol. 34: H.W. Knobloch
Higher Order Necessary Conditions
in Optimal Control Theory
V, 173 pages, 1981

Vol. 35: Global Modelling
Proceedings of the IFIP-WG 7/1 Working
Conference Dubrovnik, Yugoslavia,
Sept. 1–5, 1980
Edited by S. Krčevinac
VIII, 232 pages, 1981

Vol. 36: Stochastic Differential Systems
Proceedings of the 3rd IFIP-WG 7/1
Working Conference
Visegrád, Hungary, Sept. 15–20, 1980
Edited by M. Arató, D. Vermes, A.V. Balakrishı
VI, 238 pages, 1981